Idit Chikurel
Salomon Maimon's Theory of Invention

Idit Chikurel

Salomon Maimon's Theory of Invention

Scientific Genius, Analysis and Euclidean Geometry

DE GRUYTER

ISBN 978-3-11-099680-7
E-PDF 978-3-11-069135-1
E-PUB 978-3-11-069141-2

Bibliographic information published by the Deutsche Nationalbibliothek
The Deutsche Nationalbibliothek lists this publication in the Deutsche Nationalbibliografie;
detailed bibliographic data are available on the Internet at http://dnb.dnb.de.

© 2022 Walter de Gruyter GmbH, Berlin/Boston
This volume is text- and page-identical with the hardback published in 2020.
Printing and binding: CPI books GmbH, Leck

www.degruyter.com

Acknowledgements

This work is based on my dissertation (Tel Aviv University, 2019) and could not have come to light without the help of many, to whom I am most grateful. First and foremost, I would like to express my gratitude to my advisor, Prof. Gideon Freudenthal for his insightful guidance.

I am indebted to The Cohn Institute for the History and Philosophy of Science and Ideas and to the School of Philosophy, Linguistics and Science Studies, Tel Aviv University for both the spiritual and material support that has made this work possible. I am also grateful for the support I was fortunate to receive from the Dean of the Lester and Sally Entin Faculty of Humanities, Tel Aviv University, the Office of International Academic Affairs, Ignatz Bubis Jewish Studies Research Grant, Yaniv Foundation, the Chaim Rosenberg School of Jewish Studies and Leo Baeck Institute for the Study of German and Central European Jewry.

I wish to thank Michael N. Fried and Yaron Senderowicz for their careful reading and their enlightening comments. I owe whatever little understanding I have of Greek mathematical analysis to David Rabouin and Alain Bernard and I thank them for their guidance and time. My gratitude is extended also to Yosef Schwarz for sharing his knowledge and offering support and Rivka Feldhay who kindly invited me to join the research group *Migrating Knowledge* at the Minerva Humanities Center.

Last but not least, I wholeheartedly thank my parents, Miriam and Haim, for their boundless support and Stefan, for the constant encouragement.

Contents

Abbreviations —— IX

Introduction —— 1

Chapter 1: The Genius and the Methodical Inventor —— 6
1.1 The Genius and the Methodical Inventor —— 6
1.2 *Geniezeit*: Genius in the 18th Century —— 12
1.3 The Methodical Inventor in the 17th Century —— 21
1.4 The Objective Reality of the Genius and the Methodical Inventor —— 25

Chapter 2: An Art of Finding Arguments —— 27
2.1 Invention and Discovery —— 27
2.2 An Art of Finding Arguments —— 31
2.3 The Given —— 41

Chapter 3: Invention, Analysis and Synthesis —— 47
3.1 A General Definition of Invention as Based on Syllogisms: Analysis and Synthesis —— 47
3.2 Two Meanings of Invention as Logical Analysis —— 49
3.3 Categorical and Hypothetical Judgments —— 53
3.4 Two Kinds of Analysis in Mathematics —— 55
3.5 Two Kinds of Analysis in Philosophy: Analysis of the Concept and Analysis of the Object —— 59
3.5.1 Analysis of the object: educing a predicate immediately from the object – the case of educing three angles —— 61
3.5.2 Analysis of the object: educing a predicate indirectly by using demonstration – the case of the Pythagorean Theorem —— 65
3.5.3 The method of analysis of the object: transformation of the given object – the case of *Elements* I.5 —— 67
3.6 Invention as Synthesis —— 68
3.7 How are Synthetic *a priori* or Ampliative Analytic Judgments Possible? —— 72
3.8 Analysis and Discovery, Synthesis and Invention —— 76

Chapter 4: Methods of Invention —— 80
4.1 Seven Kinds of Analysis —— 83
4.1.1 Analysis of the Conditions of a Problem or Proposition —— 84
4.1.1.1 True and Pseudo-Conditions —— 85
4.1.1.2 Conditions of Possibility of the Solution —— 89
4.1.1.3 Conditions of the Problem —— 89
4.1.2 Analysis of a Complex Problem or Complex Proposition into Simple Ones —— 92
4.1.3 Analysis of the Cases of a Problem or a Proposition —— 101
4.1.4 Analysis of the Object —— 107
4.1.5 Analysis of the Cases of the Solution —— 112
4.1.6 Analysis of the Various Ways in which a Problem Can be Solved or a Proposition Proven —— 118
4.1.7 Logical Analysis —— 124
4.1.7.1 Regressive Analysis —— 126
4.1.7.2 Syllogisms —— 128
4.2 Conversion —— 132
4.3 Generalization —— 137
4.4 Assuming a Problematic Proposition as True —— 141

Conclusion —— 147

Bibliography —— 154

Index of Terms —— 163

Index of Person —— 167

Abbreviations

Autobiography	Maimon, Salomon (1888): *Solomon Maimon: An Autobiography* (1st ed. 1792). Murray J. Clark (Tr.). London: Alexander Gardner.
Baco	Maimon, Salomon (1793): *Bacons von Verulam Neues Organon. Aus dem Lateinischen übersetzt von George Wihelm Bartoldy. Mit Anmerkungen von Salomon Maimon. Zwei Bände. Mit Kupfern*. Berlin: Gottfried Carl Nauclk.
CJ	Kant, Immanuel (1987): *Critique of Judgment*. (1st ed. 1790). Werner S. Pluhar (Tr.) Indianapolis/Cambridge: Hackett Publishing Company.
CpR	Kant, Immanuel (1983): *Critique of Pure Reason*. (1st ed. 1780, 1787). Kemp Smith, Norman (Tr.) London: Macmillan, (2nd ed.).
CSM	Descartes, René (1985): *The Philosophical Writings of Descartes*, Vol. 1. John Cottingham/Robert Stoothoff/Dugald Murdoch. (Trs.). Cambridge: Cambridge University Press.
Das Genie	Maimon, Salomon (1971): "Das Genie und der methodische Erfinder (aus: Berlinische Monatsschrift. 1795. BD XXVI. S. 362–384)". In: Valerio Verra (Ed.). *Gesammelte Werke*. Vol. 6. Hildesheim: G. Olms Verlagsbuchhandlung, pp. 397–420.
Encyclopédie	Diderot, Denis/d'Alembert, Jean le Rond (Eds.) (2016): Encyclopédie, ou dictionnaire raisonné des sciences, des arts et des métiers, etc. Robert, Morrissey and Glenn, Roe (Eds.) University of Chicago: ARTFL Encyclopédie Project (Spring 2016 Edition). http://encyclopedie.uchicago.edu/.
Erfindungsmethoden	Maimon, Salomon (1976): "Erfindungsmethoden". In: "Ideen und Plane aus S. Ms hinterlassenen Papieren (ebenda, 1804, Heft II, S. 139–156)". In: Valerio Verra (Ed.): *Gesammelte Werke*. Vol. 7. Hildesheim: G. Olms Verlagsbuchhandlung, pp. 649–660.
Giv'at Hamore	Maimon, Salomon (1966): *Giv'at Hamore*, (1st ed. 1791). Samuel Hugo Bergman/Nathan Rotenstreich (Eds.). Jerusalem: The Israeli Academy of Sciences and Humanities [in Hebrew].
KdA	Maimon, Salomon (1794): *Die Kathegorien des Aristoteles, mit Anmerkungen erläutert und als Propädeutik zu einer neuen Theorie des Denkens dargestellt von Salomon Maimon*. Berlin: Ernst Felisch.
KrU	Maimon, Salomon (1797): *Kritische Untersuchungen über den menschlichen Geist oder das höhere Erkenntniß – und Willensvermögen*. Leipzig: Gerhard Fischer dem Jüngern.
Logik	Maimon, Salomon (1970): "Versuch einer neuen Logik oder Theorie des Denkens, Nebst angeängten Briefen des Philaletes an Aenesidemus (1st ed. 1794)". In: Valerio Verra (Ed.): *Gesammelte Werke*. Vol. 5, Hildesheim: G. Olms Verlagsbuchhandlung.
Nouveaux Essaies	Leibniz, Gottfried Wilhelm (1765b): "Nouveaux essaies sur l'entendement humain (1704)". In: *Oeuvres philosophiques latines & françoises de feu Mr. de Leibnitz: tirées de ses manuscrits qui se conservent dans la Bibliothèque Royale à Hanovre, et publiées par Mr. Rud. Eric Raspe, avec une préface de Mr. Kaestner, Professeur en mathématiques à Göttingen*.

	Rudolph Erich Raspe (Ed.). Amsterdam, Leipzig: Jean Schreuder, pp. 1–496.
PhW	Maimon, Salomon (1791): *Philosophisches Wörterbuch, oder Beleuchtung der Wichtigsten Gegenstände der Philosophie, in alphabetischer Ordnung. Erste Stück*. Berlin: Johann Friedrich Unger.
Prg	Maimon, Salomon (1969): "Ueber die Progressen der Philosophie veranlaßt durch die Preisfrage der königl. Akademie zu Berlin für das Jahr 1792: Was hat die Metaphisik seit Leibniz und Wolf für Progressen gemacht? (1st ed. 1793)". In: *Aetas Kantiana*. Vol. 172. Bruxelles: Culture et Civilisation.
Tr	Maimon, Salomon (2010): *Essay on Transcendental Philosophy* (1st ed. 1790). Midgley, Nick/Somers-Hall, Henry/Welchman, Alistair/Reglitz, Merten (Trs.). London, New York: Continuum International Publishing Group.
Ueber den Gebrauch	Maimon, Salomon (1971): "Ueber den Gebrauch der Philosophie zur Erweiterung der Erkenntnis. (aus: Philosophisches Journal. 1975. Bd. II, S. 1-35)." In: Valerio Verra (Ed.): *Gesammelte Werke*. Vol. 6. Hildesheim: G. Olms Verlagsbuchhandlung, pp. 362–396.

Introduction

> Whatever light I may receive, I shall always make it luminous with the light of reason. I shall never believe that I have fallen upon new truths, if it is impossible to see their connection with the truths already known to me. (*Autobiography*, p. 256–257)

Reason slowly shedding its light on connections between various truths is Maimon's guiding idea both as a private man and as a philosopher, the two almost always one and the same. The above quote wherein he discusses a possible conversion of religion reveals that the true shepherd of his life is reason – a thought that might have been taken from his philosophical works, first and foremost from his investigations of invention. For in Maimon's eyes, our goal in life as well as our quest for expansion of human knowledge can be summed up as finding unknown connections between truths.[1]

Our finite understanding differs from the infinite understanding only in degree (*Giv'at Hamore*, p. 34). Discovering connections between truths and inventing new ones, as well as discovering new attributes and inventing new concepts, are more than just productive actions of human beings – they are mirror images of the activity of creation by the infinite understanding. The inventive faculty is what separates us from animals (*Ueber den Gebrauch*, p. 5); the ability to invent is what it means to be created in the image of God.

Maimon's interest in invention was not unique. The tradition of theories of invention is long and rich and precedes Maimon. It was a common topic in his lifetime as well. This tradition includes some controversies; the most notable of which are whether a theory deals with invention or discovery and whether invention or discovery is analytic or synthetic. Maimon's response to these controversies was to eliminate the questions by including all possible answers in his work: his theory of invention includes both invention and discovery and both analysis and synthesis. In my work, I endeavor to present Maimon's outlook on these questions and develop his ideas and methods one step further.

Some of the works in the *ars inveniendi* tradition are closely related to mathematical innovations. Maimon also chooses mathematics to be the exemplar on which to apply his methods of invention. His work is concentrated on arriving at new mathematical, rather than philosophical knowledge, maintaining that mathematical inventions are real and not merely formal like those in philosophy.

---

[1] Shedding the light of rationality includes even the more obscure areas of human experience, such as those found in Maimon's attempts to describe parapsychological phenomena in a methodical manner (see Bergman 1967, p. 293).

However, unlike some of his predecessors such as Descartes or Leibniz, Maimon does not offer new mathematical objects or even completely new methods. Rather, he improves the existing methods of Euclidean geometry. His primary purpose in developing a theory of invention is to find a secure method for solving problems and finding new truths. He extracts methods from given mathematical inventions using examples mainly from Book I of Euclid's *Elements*. Moreover, when writing on analysis he comments on Johann Christoph Schwab's introduction to Euclid's *Data*, a book that is also associated with theories of invention. In this book, I examine Maimon's methods and examples and apply them to new examples from both of Euclid's works.

Invention is a recurring theme in Maimon's work, from his first published book (*Tr.*, p. 70) to his posthumous *Methods of Invention* (*Erfindungsmethoden*). Even though he intended to write a book on the subject, *Perfection of the Inventive Faculty through the Study of Mathematics* (*Vervollkommnung des Erfindungsvermögens durch das Studium der Mathematik*; *Das Genie*, p. 383–384; *Ueber den Gebrauch*, p. 18),[2] he never published an entire work presenting his theory of invention. He only published two articles on the subject in 1795: *On the Use of Philosophy for Expanding Knowledge* (*Ueber den Gebrauch der Philosophie zur Erweiterung der Erkenntniß*) and *The Genius and the Methodical Inventor* (*Das Genie und der methodische Erfinder*). In both articles, he only gives the outline of his theory of invention. *Methods of Invention* (*Erfindungsmethoden*) is another work dedicated specifically to the subject of invention, but it was not published in his lifetime. We can safely assume it was written after the publication of the articles from 1795 since it contains amendments to both his notions of invention and his methods. It is highly plausible that Maimon did not complete his intended book due to the complexity of his views on the subject.[3] For instance, in some of his works he mentions some methods of invention as highly important, but then completely ignores them in the works dedicated specifically to invention. As part of my work, I offer an explanation for the absence of such methods. Another difficulty that could have impeded Maimon's publication of a full, systematic book is the plurality of notions he attributes to invention and the relations

---

[2] A year before the publications of his articles on the subject, in 1794, he mentions in a letter to Goethe that the future name of his work is *On Scientific Genius, or the Ability to Invent as a Part of a Theory of Invention* (*Ueber das wissenschaftliche Genie, oder das Erfindungsvermögen, als Beitrag zu einer Theorie der Erfindung*; Schulz 1954, p. 285).

[3] Taking into consideration that Maimon was accustomed to writing in the form of commentary rather than presenting a complete system (see Freudenthal 2004), it becomes even more probable that difficulty in systematization was a major reason that prevented him from completing this work in his lifetime.

between them. In the articles from 1795 he mentions that invention is based on analysis and presents two kinds of analysis: logical analysis and analysis in a broader sense which includes intuition as its ground (alongside the principle of contradiction). In *Methods of Invention* he adds a third notion of invention: synthesis. On the one hand, he explicitly declares that logic is insufficient for invention and this insufficiency is what calls us to look for a theory of invention. On the other hand, it seems that in essence, Maimon views invention in logical terms since his most general definition of invention is one of forming syllogisms. His methods of invention are only meant, according to him, to find propositions that can be used as either premises or conclusions in syllogisms. My application of the outlines of Maimon's theory and methods on geometrical examples is meant to put them to the test and consequently offer amendments of these methods, where necessary.

The first chapter of this work describes Maimon's unique views on the genius and the methodical inventor. He believes it is important to study the work of the methodical inventor rather than that of the genius since the methodical inventor gives an account of every step taken to arrive at his conclusions. This stance was unique among thinkers of his time, a period known as *Geniezeit*. Resonating with popular ideas from the preceding century and guided by the idea of "the light of reason," Maimon argues that both the genius and the methodical inventor arrive at new knowledge in a similar way: both search for premises, middle terms and conclusions. The only difference is that the genius is not given any propositions to serve as premises and he is also unconscious of the way in which he operates. This chapter describes different aspects of the genius and the methodical inventor such as their relation to the imitator, their relation to originality and their work in the sciences. I trace the various influences on Maimon's work which include not only adopting features of his general view from other authors, but at times also specific examples and expressions, all accommodated to fit his exclusive outlook on the subject.

The second chapter reviews features of Maimon's theory of invention. The first part is dedicated to the difference between invention and discovery. Following this is a presentation of the definition of invention in general as finding propositions to be used as premises or conclusions. This definition includes both invention and discovery. Due to the plurality of definitions of *invention* presented by Maimon, this section deals only with the general definition. A more detailed discussion is presented in the following chapter. The second section of Chapter 2 attempts to explain the reasons for defining invention in general as the formulation of syllogisms. It speaks of possible influences as well as Maimon's surprising exclusion of methods of invention that are not presented in his published works on invention, but do appear in other works. As part of a discussion on the importance of actuality in his theory, I refer to his general conception of geometry. Since for Maimon all invention begins with something given, I also examine the concept of *given*. I find that Maimon's work on invention sheds new light on the use of this concept in his philosophy in general.

In the third chapter I examine Maimon's definitions of invention as analysis and synthesis. Following an examination of his two notions of invention as logical analysis, I suggest that his notion of analysis should be divided into analysis in the narrow sense (that which is grounded on the principle of contradiction alone), and analysis in the broad sense (that which is also grounded on intuition and not solely on the principle of contradiction). The latter has different forms in mathematics and in philosophy, hence, these are addressed separately. The mathematical notion of analysis in the broader sense is based on methods of analysis for problem-solving. In order to better understand the philosophical notion of ampliative analysis, the different kinds of Maimon's analysis of the object (as presented in his *Logic*) is discussed. In this form of analysis, a new predicate is educed from the object of the subject and not from the concept. Then follows a brief discussion on the third meaning, that of invention as synthesis. Although Maimon's principle of determinability is not an explicit part of his own discussion on invention, I chose to include a short discussion of the three criteria entailed in this principle along with suggestions for its amendment. The last two sections of the chapter are dedicated to the difference between ampliative analytic propositions and synthetic propositions, based on the difference of relation of determinability. It addresses Maimon's version of Kant's question of the possibility of synthetic *a priori* judgments that includes ampliative analytic propositions. The chapter concludes with a discussion of the similarity between the definitions of discovery and analysis of the object, and between the definitions of invention and synthesis.

The core of Maimon's theory of invention is the methods he presents. The last chapter contains a close examination of each of his ten methods included in his works dedicated to this subject. These methods are not completely new but rather extracted from known methods already in use by mathematicians. I suggest that in this Maimon was influenced by Proclus either directly or indirectly. I argue that Maimon's contribution to the tradition of *ars inveniendi* and to the body of mathematical knowledge is not in presenting new proofs and solutions, but in the new articulations and amendments of these methods. Maimon's methods and examples are all taken from Euclidean geometry. I follow Maimon's examples with my own, thus discovering new aspects of the methods. In the case of the seventh method of logical analysis, for instance, I suggest a redefinition of this method.

Maimon presents seven kinds of analysis that are methods of invention, the sixth being "analysis of the various ways in which a problem can be solved or a proposition proven." It expresses the idea that all truths are connected. What could have been regarded as a trivial suggestion – to prove a proposition in more than one way – is seen in new light when we take into account that exploring relations between truths is at the heart of Maimon's vocation. For Maimon, invention is not only a popular topic in philosophy, but also a guiding principle of life.

Chapter 1:
The Genius and the Methodical Inventor

Maimon wrote on the subject of genius and the methodical inventor on many occasions, devoting an article to the subject in 1795: *The Genius and the Methodical Inventor*. This chapter examines Maimon's views on this matter showing how his interest in the methodical inventor is exceptional in light of the interest in genius during his time – a time also known as *Geniezeit*.

1.1 The Genius and the Methodical Inventor

Before discussing the various influences on Maimon's standpoint on genius and methodical invention, the main features of his position should be addressed. Maimon's interest in the genius is much more limited than his interest in the methodical inventor, since he claims that one is bestowed with genius and there is nothing one can do in order to turn into a genius. Yet much can be done to become a methodical inventor and improve the methods of invention. One cannot choose to be a genius, but one can consciously choose to become a methodical inventor, who knows both how to invent and how to articulate the manner in which he arrives at his inventions. The genius only knows how to operate by way of invention without knowing the rules he follows.[4]

In the beginning of *The Genius and the Methodical Inventor*, Maimon claims that the genius and the methodical inventor are much similar than is customary to think (*Das Genie*, p. 362–363). He first mentions the differences between the genius and the methodical inventor, differences which he considers to be subjec-

---

[4] The methodical inventor "knows how" to invent and "knows that", i.e., knows the rules by which he invents, whereas the genius only "knows how." In discussing the notions of "knowing that" and "knowing how" in *The Concept of Mind* (1949), Gilbert Ryle regards invention only as "knowing how": "First, there are many classes of performances in which intelligence is displayed, but the rules or criteria of which are unformulated. The wit, when challenged to cite the maxims, or canons, by which he constructs and appreciates jokes, is unable to answer. He knows how to make good jokes and how to detect bad ones, but he cannot tell us or himself any recipes for them. So the practice of humour is not a client of his theory. The canons of aesthetic taste, of tactful manners and of inventive technique similarly remain unpropounded without impediment to the intelligent exercise of those gifts" (Ryle 1951, p. 30). Maimon's project of developing a theory of invention is based on formulating rules for inventing. His goal is to break down "knowing how" into "knowing that" – meaning, expressing rules that may assist inventors' work.

tive rather than objective. The most important difference is in respect to their knowledge (or lack thereof) of the rules according to which they act:

> The *inventive faculty*, just like any other of which we have a definite concept, must be subjected to its own unique rules, according to which it must act. Once these are found and represented in a distinct and definite way, it will appear that to deserve this high title the *genius*, just as the *methodical inventive faculty*, is subjected to the same rules: with the sole difference (which is not to the advantage of genius) that the methodical inventive faculty acts *with*, whereas genius *without*, knowledge of these rules; such that they differ from one another not in an *objective*, but only in a *subjective* way of acting. The methodical inventive faculty also provides a *touchstone* of genius; because that which cannot be methodically invented, is to be held not as a work of genius, but rather as a work of *chance*. (*Das Genie*, p. 365–366)

Both the genius and the methodical inventor act according to the rules of the inventive faculty, yet only the latter knows these rules. Both oppose acting by "mere chance" with Maimon stressing that the actions of a genius while not guided by method, are at the same time not arbitrary. Maimon's remark about the methodical inventor as being a touchstone of genius emphasizes the strong relation between the two. Both originate from the same faculty, thus illustrating his claim that their difference is only subjective.

Maimon focuses the discussion on the methodical inventor and not on the genius because of the lack of knowledge of the rules demonstrated by the genius:

> Now, on what might the invention [*das Erfinden*] or discovery [*das Finden*] be based? "On *Genius!*" I hear the resounding answer. This would certainly be a very easy way to sneak out of the matter without losing face. However, I would further like to ask, what is *genius*? "The faculty to invent." Good! In respect to its *generation*, invention is here thus explained: invention is based on the faculty to invent! A faculty that is *determined* and *distinguished* from all other faculties, by the fact that the laws according to which it functions are unknown, not only to this faculty, but also in general. But such a faculty is *not a determined* faculty but the concept of a *faculty in general*. (*Das Genie*, p. 364–365)[5]

Maimon criticizes two claims that were apparently common in his time: the first is that invention is based on genius and that genius itself is the inventive faculty. He points out the tautology in such an explanation. The second claim is that what distinguishes the inventive faculty from other faculties is that its laws are unknown. Maimon's response to such a claim is that this criterion cannot serve to distinguish the inventive faculty from any other since any faculty as

---

5 See also: *Ueber den Gebrauch*, p. 12.

such is characterized as one whose laws are unknown. For instance, elsewhere he mentions that the faculty of intuition [*Anschauungs-Vermögen*] "certainly conforms to rules but does not comprehend rules [...]" (*Tr.*, p. 34–35), thus indicating that the inventive faculty is not distinguished from other faculties in this respect.

The methodical inventor and the genius differ also in the manner in which they arrive at new knowledge. Following a rule consciously (the methodical inventor) or unconsciously (the genius), they establish different starting points in the procedure for arriving at unknown truths. Maimon claims that inventing a new unknown truth from known truths requires the sorting of already acquired truths in one's cognition. Those known truths can serve as premises and the unknown truth serves as the sought conclusion. The methodical inventor is given the premises that lead to the unknown conclusion or is given the conclusion and is to find the premises that lead to it. However, neither conclusion nor premises are given to the genius and he himself selects the truths that lead to the yet unknown conclusion (*Das Genie*, p. 366–367).[6]

The action of the genius, writes Maimon, "comes close to inspiration" (*Das Genie*, p. 367), with the word "inspiration" applying to solving problems in geometry as well as to writing poetry. An example of such use of inspiration in geometry can be found in Newton whom Maimon calls a "genius." In *Essay on Transcendental Philosophy* (1790) Maimon writes:

> It is well known that no general method has yet been discovered to prove a geometrical theorem or to solve a problem; instead it comes down to certain tricks in drawing the so-called preparatory lines. But we can draw God knows how many of these and connect them both to each other as well as to those already drawn in a variety of ways, and yet still either fail to reach the goal or reach it only after much fruitless wandering. So it is a feature of genius,

---

6 "*To invent* a new unknown truth from the already known requires that the inventor sort out of the entire mass of truths that make up his already acquired cognition, those truths that can serve as *premises* of the *conclusion* which is the truth to be invented. Here are hence two possible cases. *Either* the truth to be invented is given as a problem or theorem, to which he should find the proof or the solution as the *middle term* in order to arrive to this *end*; and so also in the reverse: those truths particularly marked through the connection of which the truth is to be invented, which is not told to him, can be brought out. This is the business of the *methodical inventor. Or* again: neither are particular truths out of the entire mass of his knowledge marked as premises nor is he given an undetermined known truth set as an end; and he should select himself out of the mass of his knowledge those truths that lead by their connection to some unknown truth as to an *end as such*. This is the business of the *genius*. It comes close to *inspiration* since it presupposes an inexplicable *circle* in the operations of the cognitive faculty" (*Das Genie*, p. 366–367).

that is of a kind of presentiment or instinct, to see in advance that certain lines are the ones that lead straight to the desired goal. (*Tr.*, p. 368–369)

After quoting a paragraph from Newton's *Universal Arithmetic* regarding construction in solving problems in geometry, Maimon adds:

> All this is correct, but I think you have to be a Newton to be able to use this prescription. Newton's prescription for mathematical invention seems to me like Klopstock's prescription for the higher art of poetry. Lucretius is not entirely wrong to compare the discoverer with the blood hound: "*Ut canes...*" (*Tr.*, p. 370)

Maimon emphasizes the fact that on many occasions, following a method when searching for a solution to a problem does not always suffice. When solving a problem that includes a multilateral figure, we may use Newton's method of breaking down multilateral figures into triangles. While this might assist us in advancing towards the solution, it does not guarantee it. There might be a need for further geometrical constructions that are not a part of the method. Furthermore, while the method might be of assistance in solving many problems regarding multilateral figures, it could also happen that, with regard to a specific problem, the method does not assist us.

We cannot determine whether inventions are the result of a work of genius or of a method, claims Maimon, and even a great number of inventions cannot serve as a criterion for genius (*Das Genie*, p. 368). He mentions Euclid's *Elements* as a work that could have been written through "persistent study, much profoundness and acuteness, much order and method" (*Das Genie*, p. 369), but without genius. Maimon adds that "Genius is characterized above all by speed and easiness of its action" (*Das Genie*, p. 369), which is the opposite of the well-thought-out step by step manner of the methodical inventor. Inventions could arise spontaneously and without much effort, but their ordered way of presentation makes it appear as though they had arisen by using a method. In the same way, one cannot conclude that a work revealing much order and method lacks genius (*Das Genie*, p. 369).[7] It is important to mention that the "speed

---

[7] Adopting Reichenbach's notions of *context of discovery* and *context of justification* (Reichenbach 1938, p. 6–7; p. 383–384), we cannot determine that one's work is a work of a methodic inventor merely because it manifests both contexts – by presenting the new theorem with its proof. It seems that the work of a genius can manifest both as well, not only the first. For instance, there are cases in which the proof itself contains a "product of genius", such as Newton's geometrical proof mentioned by Maimon (*Tr.*, p. 368–369), where the method of breaking down multilateral figures into triangles can be considered as "an act of genius." In this aspect, Maimon is a forbear of the approach that invention is a rational action, wherein the manner in

and easiness" of the genius can also be the result of working as a methodical inventor which serves as a platform for knowledge that allows the "spark of genius." Ernst Mach mentions that "moments of genius" are the result of hard labor in mastering a body of knowledge and are a result of systematic method, be it conscious or unconscious. The result is an idea that "has poured into" the genius. He also mentions Newton as one whose genius was a result of hard labor (Mach 1986, p. 174). In Maimon's terms, the "speed and easiness" that characterize the genius' work could be a result of the hard work of acquiring the propositions and methods that will later serve him in arriving at new conclusions (even if his use of this knowledge is done unconsciously).

The methodical inventor may not have the good fortune of being "led to his inventions by lucky ideas" (*Das Genie*, p. 370) as does the genius, but he has the advantage of progressing in the same manner from the beginning until the end of his work (*Das Genie*, p. 370). Maimon also claims that "Genius is like a woman, who easily conceives but gives birth under pains" (*Das Genie*, p. 369), emphasizing that the genius might have the advantage of "speed and easiness" in arriving at his invention, but then faces great difficulty in elaborating it.[8] Genius is considered a "gift of nature" and the work of the genius cannot be learned, unlike the methodical inventor's work that can be learned and improved (*Ueber den Gebrauch*, p. 15; *Das Genie*, p. 371). Since we cannot have insight into the manner by which the genius invents, Maimon puts his major effort into discussing the methodical inventor and not the genius. To a possible claim that all that is needed for invention is genius, Maimon answers that his theory can be of much value when genius is lacking, and so "give its action the proper direction" (*Ueber den Genrauch*, p. 3–4).[9]

---

which we arrive at a new idea is not, as Popper maintained, "irrelevant to the logical analysis of scientific knowledge" (Popper 2005, p. 7), but rather that we can account for how knowledge is produced and advances (e.g. Zahar 1983).

8 A similar idea appears in Kant's *Critique of Judgment:* "For the scientists' talent lies in continuing to increase the perfection of our cognitions and of all the benefits that depend on [these], as well as in imparting that same knowledge to others; and in these respects they are far superior to those who merit the honor of being called geniuses. For the latter's art stops at some point, because a boundary is set for it beyond which it cannot be extended further" (§ 47, p. 309).

9 Kuntze writes on genius not only as a gift of nature, but also as an inborn gift of culture: genius is based on the use of freedom to connect the "fertility of spirit" with the "gift of judgment," when the "fertility of spirit" wishes to avoid submitting to objective rules. The "gift of nature" cannot be imitated, but his use of freedom with the two faculties ("fertility of spirit" and judgment) can be (Kuntze 1912, p. 451). Similarly, Kant claims that the genius' free use of his cognitive power is not to be imitated, but followed by others: "The other genius, who follows the example,

Alongside the genius and the methodical inventor appears a third character: the imitator. Maimon does not mention him by name, but by character. In Maimon's time, the imitator was known as a character opposite to that of the genius. After discussing the different processes that the genius and the methodical inventor go through in order to arrive at unknown truths, Maimon writes:

> According to this, the methodical inventor hence stands right in the middle between the genius, *who draws only on his own*, and the one who learns merely the inventions of others. The genius is to be compared with the first bold seaman, who dared embark on the ocean without help of a compass; the methodical inventor [is to be compared] with the seaman who invented the compass for the purpose of navigation; and our contemporary seaman is to be compared with a person who only learns and uses the invention of others. (*Das Genie*, p. 367–368)

The methodical inventor stands in between the genius, who is original and relies only on himself, and the imitator who relies only on the work of others. The methodical inventor relies both on the work of others as well as on his own. The genius, Maimon adds, can be unconscious of the laws that he follows in order to arrive at these truths, or perhaps, aware of these laws, is not concerned about presenting them to his readers. The methodical inventor will "search out the secret ways of the genius" (*Das Genie*, p. 370) and will present the readers his method. The imitator (not mentioned by name but only as "a third") will not invent but only use inventions without considering their way of creation (*Das Genie*, p. 369–370).[10] Maimon places the imitator in opposition not only to the genius, but also to the methodical inventor. The latter opposition stands out in light of Maimon's contemporaries who speak only of the genius and the imitator, but do not mention the methodical inventor. The next section refers to some of the important works on genius in Maimon's time, known also as *Geniezeit* (or *Genieperiode*). The absence of the methodical inventor in the discussions on genius sheds light on how important the methodical inventor is to Maimon and how innovative his perspective is in light of his time.

---

is aroused by it to a feeling of his own originality, which allows him to exercise in art his freedom from the constraint of rules, and to do so in such a way that art itself acquires a new rule by this, thus showing that the talent is exemplary" (*CJ*, § 49 p. 318).

10 In the entry *Imitation* [*Nachahmung*] in his philosophical dictionary, Maimon mentions that there are three classes of people. The first class is the imitators, the second class is the uncultivated, and the third class is the "more than cultivated": the geniuses, the originals, the self-thinkers and inventors in both the arts and the sciences. Maimon describes them as those "who can consider not only where things are but how they can and should be" (*PhW*, p. 84–85).

1.2 *Geniezeit:* Genius in the 18th Century

Until the mid-18th century, the word *genius* usually referred to a general attribute and not to individuals. For example, in the *Dictionnaire de l'Academie françoise* of 1694, all definitions of *genius* relate to the concept as a feature one has and not to something an individual is (Jaffe 1980, p. 581). The inconsistency of genius being something one has or as something one is, can also be seen in Maimon's work: In his philosophical dictionary (*Philosophisches Wörterbuch,* 1791), he claims that the French are more genius than the English (*PhW,* p. 88 – 89), yet in *On the Use of Philosophy for Expanding Knowledge* (1795) he writes: "Genius is a *gift of nature:* he who obtains it, enjoys it" (*Ueber den Gebrauch,* p. 14).

The genius as operating without knowledge of the rules is already a well-discussed topic in Maimon's time. In Diderot and d'Alembert's *Encyclopedie,* in the entry dedicated to genius in philosophy, genius is described as someone who does not have the disposition to follow his ideas further and who cannot be cultivated by rules (de Saint-Lambert, "Génie" in: *Encyclopédie,* Vol. 7, p. 582).[11] The product of genius is depicted both as insecure and as the opposite of order. As such, it clashes with philosophy that is a realm in which thinkers are "capable of order and following through with their ideas" (*Encyclopédie,* Vol. 7, p. 583). Yet, genius can also be of great value to philosophy:

> Genius hastens the progress of Philosophy by the most fortunate and least expected discoveries; it rises as does the flight of the eagle towards a bright truth, the source of a thousand truths gliding over the shy crowd of wise observers. (*Encyclopédie,* Vol. 7, p. 583)

Maimon also mentions the flight of the eagle in reference to genius. Emphasizing the elusiveness of the work of genius, Maimon quotes King Salomon on his inability to fathom four things, the flight of the eagle being the first and genius being the fourth (*Ueber den Gebrauch,* p. 15).[12]

Another recurring subject in discussions on genius in the 18th century is the question of imitation and originality. For instance, Kant claims that originality is the most important property of genius (*CJ,* § 46, p. 307– 308). In his philosophical

---

[11] I refer here only to the entry of *Génie* in philosophy and literature written by the poet Jean François de Saint-Lambert (1716 – 1803). *L'Encyclopedie* also includes entries for genius in mythology and ancient literature, in painting, and in the military art.
[12] Together with "The sign of genius in the depths of human cognition" and "the sign of the flight of the eagle in the air" one can find the sign of ships in mid-sea and the sign of a snake wriggling in rocks (*Ueber den Gebrauch,* p. 15).

dictionary, Maimon includes genius not as a separate entry but as a part of his entry on originality (*PhW*, p. 88–89). The dictionary was published one year after the publication of Kant's *Critique of Judgment* (*Kritik der Urteilskraft*, 1790). We know that Maimon had a copy of this book in 1790, but it is uncertain how much of it he had read.[13] In either case, the subject of genius and originality was well-discussed in Maimon's time.

The main influence on the increasing interest in the concept of genius in Germany at that time was Edward Young's *Conjectures on Original Composition* (1759), which was translated into German in 1760 (Jefferson 2009, p. 185–6; p. 195). Young describes originality as natural, opposing imitation which is unnatural and mechanical, and similar to Maimon's descriptions of genius as a "gift of nature."[14] Another major influence is Batteux's *Les Beaux-Arts Réduits à un Même Princip* (1746).[15] Batteux's notion of genius (in the arts) is that of a person who can produce art only by imitation, and therefore, only discovers something existing rather than inventing something new (Batteux 1746, p. 11). Thus, Batteux ties genius and imitation together. It seems that by the time Maimon wrote his dictionary and his article on genius, the term *genius* was well-known and much in use in Germany; it had already been connected to both originality and imitation. Placing genius within the entry on originality, situates Maimon on this side of the debate.

In his dictionary, Maimon does not dedicate a separate entry to genius, but refers to it in the entry *Originalgeist* (*PhW*, p. 88–89). He dedicates only a few words to genius and only in reference to originality. According to Maimon, even though originality and genius often reside together, they are distinguished from one another. Maimon defines *originality* as a particular form and the *spirit of originality* [*Originalgeist*] as a unique way of acting that is distinguishable from

---

[13] In a letter to Kant, from May 15th 1790, Maimon mentions that he has not thoroughly read *Critique of Judgment*, but only flipped through it (*Maimon an Kant*, 15 Mai 1790; in: Maimon 1970, p. 429–431).

[14] Young writes: "An *Original* may be said to be of a *vegetable* nature; it rises spontaneously from the vital root of Genius; it *grows*, it is not *made*: Imitations are often a sort of *Manufacture* wrought up by those *Mechanics*, *Art*, and *Labour*, out of pre-existent materials not their own" (Young, *Conjectures on Original Composition* 1966, p. 12; in: Jefferson 2009, p. 185).

[15] Friedrich Kluge's *Etymological Dictionary of the German Language (Etymologisches Wörterbuch der Deutschen Sprache*, 1960), mentions that the word *Génie* first appeared in German at the beginning of the 18th century and became popular in 1751 following Johann Adolf Schlegel's translation and commentary of Batteux's *Les Beaux-Arts Réduits à un Même Princip* (Kluge 1960, p. 247). Kluge also mentions that the word *Génie* was derived from Latin and introduced into Italian in the 17th century. It was in use in French sometime before that, but its dating is unspecified (Kluge 1960, p. 247).

the rest. *Genius*, however, is "an excellent level of action [*Wirkung*]" (*PhW*, p. 88–89). It seems that the main difference concerns whether the way of action is in or outside the common way of action: originality offers a manner of action outside the common realm, whereas genius is set within the common realm, exhibiting a very high level of performance. An example of such a difference could be found in Leibniz's and Newton's calculus: Leibniz's *differential* is a "work of genius": the differential is a geometrical magnitude (small as it may be), much like the infinitesimals and indivisibles presented before him by mathematicians such as Cavalieri and Torricelli, which are geometrical magnitudes with no movement involved.[16] Leibniz's improvement of the concept shows an "excellent level of acting" within existing discourse. Newton's *fluxion*, on the other hand, is original since motion is added to the concept of the infinitesimal,[17] thus showing a "way of action" that is different from the common concept and discourse.

Maimon claims that originality and genius are analogous to various animal instincts. Each animal species has its own instinct, its way of acting, a skill that requires no practice to perform it. Maimon adds that the French are more genius than the English, but the English are more original. He also mentions Montesquieu as one who was more genius, and Rousseau as more original-spirited (*PhW*, p. 88–89).[18] This notion of the English as more "original-spirited"

---

[16] We have several pieces of evidence that Leibniz regarded differentials as geometrical entities. In Leibniz's first written work on the calculus, *A New Method for Maxima and Minima as well as tangents, which is impeded neither by fractional nor irrational quantities, and a remarkable type of calculus for them* (published in *Acta Eruditorum* in 1684), he defines dx and dv as straight lines. He also adds a graph of dx as a line for illustration (Leibniz 1983, p. 4). Further evidence of Leibniz's notion of differentials as geometrical entities can be found in his letter to Nieuwentijt (1694): "Therefore I accept not only infinitely small lines, such as dx, dy, as true quantities in their own sort, but also their squares or rectangles, such as $dxdx$, $dydy$, $dxdy$. And I accept cubes and other higher powers and products as well, primarily because I have found these useful for reasoning and invention" (*Mathematische Schriften V.* p. 322; in: Bos 1974, p. 64). Even though Leibniz also accepted higher-order differentials, one can see by his use of terms such as "lines," rectangles" and "cubes" that the main notion of a differential is geometric (unlike Newton's fluxion defined as motion). Leibniz was influenced by Cavalieri (Beeley 2008, p. 44), who called the indivisibles of a figure by the name "all the lines" (*Geometria*, p. 114: "indivisiblia. s. omnes lineas fugurae"; in: Andersen 1985, p. 301).

[17] In *The Method of Fluxions and Infinite Series* (written in 1671, first published in 1736) Newton writes: "Now those Quantities which I consider as gradually and indefinitely increasing, I shall hereafter call *Fluents*, or *Flowing Quantities,* [...] And the Velocities by which every Fluent is increased by its generating Motion (which I may call *Fluxions*, or simply Velocities or Celerities) [...]" (§ 60, p. 20).

[18] Rousseau himself wrote on *Génie* in his *Dictionnaire de Musique* (1768), explaining to the "young artist" that there is no point in searching for what genius is: if he has it, he will

than the French, appears also in Kant's *Phillippi Logic* (1772).[19] It is unclear whether Maimon was influenced by Kant's approach to the English as having original spirit, or whether both Kant and Maimon were influenced by a third party or the general public view, yet it is clear that Maimon was influenced by the discourse on genius and originality that was taking place during his time.

Originality, according to Kant, is the most important property of genius. Genius is a talent for producing things without being given the rule, and therefore, cannot be learned (*CJ*, § 46, p. 307–8).[20] Just as it appears in Maimon's work (and in the works of others before him), Kant also emphasizes the non-methodical nature of genius:

---

sense it himself, and if he doesn't, he will never know it (p. 227). The "touchstone" for genius as a feeling one has (or does not have) appears again at the end of the entry: Rousseau sends the reader to visit Naples and its leading musicians and composers. If the reader still asks what genius is after such visit, Rousseau considers him as "Homme volgaire," asking him not to profane the sublime word *genius* (p. 228).

19 "The English, original spirits, who write without imitating others, have often recited adverse and false propositions only in order to be original" (*Phillippi Logic* (1772), in: Martin 1997, p. 577).

Tonelli mentions that Kant rarely uses the term *genius* before 1770 and when writing (in 1764) about "the Italian genius" in comparison to that of other nations (II, 244), *genius* usually means the English and French notion "génie du peuple." *Genius* also appears in the meaning of a "power of the mind" (Tonelli 1966, p. 109).

20 According to Boehm, Kant's notion of genius as something original and meaningful which can serve as a model and a standard of judgment has much to do with his change of the concept of enlightenment. Enlightenment, Boehm claims, is not merely "rejecting the guidance of another" (Boehm 2013, p. 164), but rather is similar to genius since "they both share the awkward property that following them adequately depends on not following them" (Boehm 2013, p. 172). As the work of a genius inspires one to produce other original works of genius (which will naturally be different from the inspiring work because even the artist himself cannot give an account of the steps to be taken in order to produce such a work, i.e., his process of creation lacks method and use of concepts), so the enlightened person is to be inspired by others to think for himself.

Boehm connects the rising of enlightenment to the rejection of prophecy, which itself is "intimately linked to a degradation of the faculty of imagination" (Boehm 2013, p. 150–151). The low status of the faculty of imagination, as opposed to the high status of the faculty of understanding is a prominent characteristic of Maimon's philosophy. Therefore, it is no surprise to see that Maimon's notions of genius and prophecy are both based on reason. As mentioned above, the genius works according to rules (even if unconsciously). As for prophecy, Maimon writes in his autobiography that what is revealed to the prophets is not the absolutely Perfect Being, since the Perfect Being reveals himself to reason as an idea and his relations to other natural objects are founded by the laws of nature (*Autobiography*, p. 178). Similarly, with regard to supernatural laws, Bergman mentions that for Maimon, "there might be laws which *until now* are unknown and which are therefore *with respect to us* supernatural laws" (Bergman 1967, p. 290).

Genius itself cannot describe or indicate scientifically how it brings about its products, and it is rather as *nature* that it gives the rule. That is why, if an author owes a product to his genius, he himself does not know how he came by the ideas for it; nor is it in his power [*Gewalt*] to devise such products at his pleasure, or by following a plan, and to communicate [his procedure] to others in precepts that would enable them to bring about like products. (Indeed, that is presumably why the word genius is derived from [Latin] *genius*, [which means] the guardian and guiding spirit that each person is given as his own at birth and to whose inspiration [*Eingebung*] those original ideas are due). (*CJ*, § 46, p. 308)

Unlike Maimon, Kant claims that genius is necessary for art but not for science (*CJ*, § 46, p. 308–9). According to Kant, genius is the ability to exhibit aesthetic ideas (*CJ*, § 57, p. 344). An aesthetic idea is an intuition which does not have a concept (unlike a rational idea, that is a concept without an intuition) (*CJ*, § 57, p. 342). This stance is quite different from Maimon's, which asserts that the actions of the genius can, in essence, be articulated as rules and are based on concepts. Kant states that "fine art, as such, must be regarded as a product of genius rather than of understanding and science, and hence as getting its rule through *aesthetic* ideas, which are essentially distinct from rational ideas of determinate purposes" (*CJ*, § 58, p. 351). He adds that art, as the product of genius, receives those rules not from a purpose but from the nature of the subject. The beautiful is not judged according to concepts, but according to imagination (*CJ*, § 57, p. 344). Art presupposes rules, yet we cannot use these rules in a conceptual manner in order to make judgments about beauty (*CJ*, § 46, p. 307).[21] Rules in fine art cannot be presented as a formula, since judgments

---

21 Kant states that "For every art presupposes rules, which serve as the foundation on which a product, if it is to be called artistic, is thought of as possible in the first place. On the other hand, the concept of fine art does not permit a judgment about the beauty of its product to be derived from any rule whatsoever that has a *concept* as its determining basis, i.e., the judgment must not be based on a concept of the way in which the product is possible. Hence fine art cannot itself devise the rule by which it is to bring about its product. Since, however, a product can never be called art unless it is preceded by a rule, it must be nature in the subject (and through the attunement of his powers) that gives the rule to art; in other words, fine art is possible only as the product of genius" (*CJ*, § 46, p. 307). Elsewhere, he refers to the reason he excludes the fine arts from the sciences: "There is no science of the beautiful [*das Schöne*], but only critique; and there is no fine [*schön*] science, but only fine art. For in a science of the beautiful, whether or not something should be considered beautiful would have to be decided scientifically, i.e., through bases of proof, so that if a judgment about beauty belonged to science then it would not be a judgment of taste" (*CJ*, § 44, p. 305). Fine arts are not to be based on proofs since they are based on aesthetic ideas and not on rational ideas. But this does not mean that in the production of art there is no involvement of rules other than those which originate from the nature of the subject. It is important for Kant to emphasize that the product of genius is not a product of chance, but is made with a purpose. He asserts that the production of a genius artifact involves

on the beautiful cannot be determined by concepts. The rule is abstracted from the artistic product, in a manner Kant himself refers to as "difficult to explain" and which is summed up as "the artist's ideas arouse similar ideas in his apprentice," assuming the apprentice is gifted enough to be affected by the artistic product (*CJ*, § 47, p. 309–310). Kant's approach to the rules prescribed by the genius is quite different from Maimon's. Maimon's view of the genius is as one that acts according to a rule (even if without its knowledge) and that this rule is based on concepts and can be presented by the methodical inventor as well. In essence, according to Maimon, both the genius and the methodical inventor work according to the same rule (*Das Genie*, p. 365–366). Also, it is not always possible to judge whether a work (i.e. product) was made by a genius or a methodical inventor (*Das Genie*, p. 368). In Kant's view, on the other hand, the rule of the genius is not based on concepts but (mostly) on the faculty of imagination, which makes it difficult, almost impossible, for the methodical inventor to present this rule used by the genius. Kant also regards the products of the genius and those of the methodical inventor to be of a different kind. Kant claims that "Nature, through genius, prescribes the rule not to science but to art, and this also only insofar as the art is to be fine art" (*CJ*, § 46, p. 308). This is unlike Maimon's notion that genius is found in the sciences as well.

Kant is very clear on his position in the "originality vs. imitation" debate when he attributes the main role to originality. He is also very decisive concerning the absence of imitation in genius: "On this point everyone agrees: that genius must be considered the very opposite of a *spirit of imitation*" (*CJ*, § 47, p. 308). Maimon is of the same opinion. We should note that Kant does not mention the methodical inventor, the heart of Maimon's interest and the reason for his article on the genius. Also, in light of the popular use of the word *Génie* in Germany, which was probably influenced by Batteux, it is likely that Kant addresses the subject of imitation and genius in reference to Batteux's work.[22]

Kant was also influenced by Alexander Gerard's *Essay on Genius* which was published in 1774.[23] He refers to Gerard several times and it seems that Kant responds to Gerard's occupation with scientific genius as well as genius in the arts

---

rules which originate in the mechanical arts. Kant refers to these rules as "academic correctness" (*CJ*, § 47, p. 310).
22 Kant mentions Batteux as a popular figure of authority concerning matters of taste: "If someone reads me his poem, or takes me to a play that in the end I simply cannot find it to my taste, then let him adduce *Batteux* or *Lessing* to prove that his poem is beautiful, or [bring in] still older and more famous critics of taste with all the rules they have laid down" (*CJ* ,§ 33, p. 284).
23 Gerard's *Essay on Genius* was translated into German in 1776 by Christian Garve. Garve wrote a famous critic on Kant's *Critique of Pure Reason* which is known to have influenced Kant.

(Guyer 2011, p. 59). Unlike Kant, Maimon refers to the subject of the genius mainly in the context of science and not art. For example, Maimon's article *The Genius and the Methodical Inventor* focuses on scientific genius, more specifically, genius in mathematics. However, he does not mention the genius in the arts.[24] His focus on science alone stands out in light of other works on the genius that were published at that time, which mostly dealt with aesthetics. Kant not only relates to genius as "the innate mental predisposition (*ingenium*) *through which* nature gives the rule to art" (*CJ*, § 46, p. 307) and to the fine arts as they "must necessarily be considered arts of *genius*" (*CJ*, § 46, p. 307), but he also denies any role to be played by genius in science: "genius is a talent for art, not for science, where we must start from distinctly known rules that determine the procedure we must use in it" (*CJ*, § 46, p. 317).[25] Kant's and Maimon's approaches are not only different with respect to the role of the genius in the arts and sciences, they also differ with respect to the methodical inventor. For Maimon, the main difference between the genius and the methodical inventor lies in the ability to trace the rules they operate by, when in fact they share the same rules, only differing in the knowledge of those rules (*Das Genie*, p. 365). Kant, on the other hand emphasizes that the differences between the genius and the methodical inventor are much greater and assigns each to a separate "field of knowledge" so that the two share almost no common features. While Kant does not mention the methodical inventor, it is implied that he assigns the sciences to the methodical inventor, whereas the genius is placed in the realm of the arts. Kant does not say so in so many words, but it seems that Kant's suggestion that the way to decide whether something is a work of a genius or of a methodical inventor (whether in the arts or the sciences) lies in the ability to imitate the work:

> Thus one can indeed learn everything that *Newton* has set forth in his immortal work on the principles of natural philosophy, however great a mind was needed to make such discoveries; but one cannot learn to write inspired poetry, however elaborate all the precepts of this art may be, and however superb its models. The reason for this is that Newton could show how he took every one of the steps he had to take in order to get from the first elements of geometry to his great and profound discoveries; he could show this not only to himself but to everyone else as well, in an intuiteve[ly clear] way, allowing others to follow.

---

[24] As mentioned in the introduction, Maimon intended to name his book on the theory of invention as *On Scientific Genius, or the Ability to Invent as a Part of a Theory of Invention* (Schulz 1954, p. 285), stressing the fact that the genius is a scientist (and not an artist).
[25] Not only does genius not supply the rule for science, nature also does not take any part in it: "Nature, through genius, prescribes the rule not to science but to art, and this also only insofar as the art is to be fine art" (*CJ*, § 46, p. 308).

But no *Homer* or *Wieland* can show how his ideas, rich in fancy and yet also in thought, arise and meet in his mind; the reason is that he himself does not know, and hence also cannot teach it to anyone else. In scientific matters, therefore, the greatest discoverer differs from the most arduous imitator and apprentice only in degree, whereas he differs in kind from someone whom nature has endowed for fine art. (*CJ*, § 47, p. 308–309)[26]

In Kant's view, the difference between the imitator and the methodical inventor is only a difference in degree. Both are different from the genius. For Kant, the difference between the genius and the methodical inventor is objective, unlike Maimon's view that the difference between the two is only subjective (*Das Genie*, p. 365). As well, Maimon sets the methodical inventor between the genius and the imitator, unlike Kant whose division posits the methodical inventor on the same side as the imitator, opposing the genius. Remembering Maimon's words: "Newton's prescription for mathematical invention seems to me like Klopstock's prescription for the higher art of poetry" (*Tr.*, p. 370),[27] shows that

---

[26] This is not to say that Kant has no admiration to the great minds of science: "But saying this does not disparage those great men, to whom the human race owes so much, in contrast to those whom nature has favored with a talent for fine art. For the scientists' talent lies in continuing to increase the perfection of our cognitions and of all the benefits that depend on [these], as well as in imparting that same knowledge to others; and in these respects, they are far superior to those who merit the honor of being called geniuses. For the latter's art stops at some point, because a boundary is set for it beyond which it cannot be extended further. Moreover, the artist's skill cannot be communicated but must be conferred directly on each person by the hand of nature. And so it dies with him, until some day nature again endows someone else in the same way, someone who needs nothing but an example in order to put the talent of which he is conscious to work in a similar way" (*CJ*, § 47, p. 309).

Kant believes that poetry, in contrast to science, cannot be learned through examples and rules: "Among all the arts *poetry* holds the highest rank. (It owes its origin almost entirely to genius and is least open to guidance by precept or examples)" (*CJ*, § 53, p. 326). His approach to poetry and art, as opposing method and order, is too strict in light of complex reality: In Jochen Schmidt's *Die Geschichte des Genie-Gedankens in der deutschen Literatur, Philosophie und Politik 1750–1945* (1985) on the development of the concept of genius, Schmidt dedicates a short segment to the influence Descartes' method had on the development of rules in poetry. It seems that already by 1630 the rules for poetry that were accepted in France were influenced by Descartes' method and his mathematical-logical structure as presented both in his *Mathesis universalis* and his *Regulae* (1628). Schmidt claims that it is the importance of logical-deductive thought in Descartes' works that inspired poets to establish rules for their art and gave the poetic rules their high-status (Schmidt 1985, Vol. 1, p. 22–23).

[27] Maimon's notion of the genius resembles Novalis' view that although the genius acts according to rules, a genius does not follow them consciously. Thus, one cannot simply follow certain rules in order to become a genius. However, unlike Maimon, Novalis compares the genius to an artist and calls for the "poeticization" of philosophy, believing it will help in tracking the rules by which the genius works (Thielke 2015, p. 10).

he relates to Newton as both a methodical inventor and a genius, and compares him to an artist as well, since he does not make Kant's harsh distinctions between the genius and the methodical inventor and between art and science. Additionally, in Kant's opinion, imitating Newton's actions equals "following him", whereas Maimon is quite aware that studying Newton's geometric constructions does not mean one will be able to perform such acts of genius himself.

As mentioned before, Kant's emphasis that genius takes no part in science is a response to Gerard's notion of genius as playing an active role in both science and art. In his *An Essay on Genius* (1774), Gerard writes:[28]

> The ends to which Genius may be adapted, are reducible to two; the discovery of *truth*, and the production of *beauty.* The former belongs to the *sciences*, the latter to the *arts*. Genius is, then, the power of invention, either in science or in the arts, either of truth or of beauty. (Part. III., Sect. I., p. 317)

For Gerard, Genius is not only an essential part of invention, but also is the inventive power itself: "Genius is properly the faculty of invention; by means of which a man is qualified for making new discoveries in science, or for producing original works of art" (Gerard, 1774, Part I., Sect. I, p. 8). Perhaps the "resounding answer" that Maimon hears when asking on what invention is based – "On genius!" (*Das Genie*, p. 364), is in fact Gerard's idea. Gerard also states that the man of method is not an inventor, but only an imitator. Only the genius is an inventor. Once again, Maimon's approach to the methodical inventor stands out from the common view of his time.

By writing on scientific genius, Maimon was no different from Gerard or other thinkers of his time. However, the examination of Gerard's work brings out once again the innovation of Maimon's focus on the methodical inventor as an inventor at all. Maimon's interest in the methodical inventor was outstanding in light of 18th century thought, but it was a very common topic in the 17th century.

---

28 According to Larsen, in the 18th century there was a decline in the idea of invention as dependent upon genius. Bacon's scientific method outshined the common notion of *invention*, that was a part of the *topoi* (the rhetorical field which is the source of the term *inventio*). It was Gerard's *Essay on Genius* that brought back the notion of invention as the work of the genius (Larsen 1993, p. 181).

1.3 The Methodical Inventor in the 17th Century

The 17th century saw a rise of interest in the methodical inventor and his "secure methods." Searching for a secure method was a recurring motif in the works of philosophers who greatly influenced Maimon, such as Bacon and Leibniz. Maimon defines inventing as "to bring out *unknown truths out of known* truths following *secure methods*" (*Das Genie*, p. 363). It is safe to presume that Maimon's emphasis on the importance and advantages of a secure method, a recurring theme also in his work, was influenced by these philosophers.

Maimon begins his article *The Genius and the Methodical Inventor* with a citation from Bacon's preface of *Great Instauration* (1620) (*Das Genie*, p. 362).[29] In this preface Bacon writes:

> We must guide our steps by a clue, and the whole path, from the very first perceptions of our senses, must be secured by a determined method [...] But, as in former ages, when men at sea used only to steer by their observations of the stars, they were indeed enabled to coast the shores of the Continent, or some small and inland seas; but before they could traverse the ocean and discover the regions of the new world, it was necessary that the use of the compass, a more trusty and certain guide on their voyage, should be first known. (Bacon 1859, p. 336)

A secure method must be secure from the beginning, working its way safely, step by step. This requirement appears in Maimon's work as well (*Das Genie*, p. 373; p. 370). It is plausible that Maimon's example of the use of a compass as a "secure method" for seamen in their navigations (*Das Genie*, p. 367) was influenced by this text.[30]

Two features in Maimon's work on the methodical inventor are method and order. Receiving high importance in 16th and 17th century English philosophy, and in 17th century French philosophy, they are central to the philosophical thought of the time (Wallace 1973, p. 272; p. 248). Bacon, for instance, uses the image of

---

29 Maimon cites the Latin of the following sentence: "It appears to me that men know not either their acquirements nor their powers, and trust too much to the former, and too little to the latter." (Bacon 1859, p. 334).
30 In 1793, Maimon published a commentary on Bacon's *Novum Organum*, the second part of *Great Instauration*, accompanying Bartoldy's German translation (*Bacons von Verulam Neues Organon*, 1793). It includes two appendices: *A Short Exposition of Philosophical Systems* (*Kurze Darstellung Philosophisher Systeme*) in which Maimon writes on philosophical systems from Thales to Pyrrho, including discussions such as Xenophon as a Leibnizian, and Pyrrho's system in light of the Kantian system; as well as *A Short Exposition of Mathematical Inventions* (*Kurze Darstellung Mathematischer Erfindungen*), which follows Montucla's *Histoire des mathématiques* (1758) and in which Maimon presents important known mathematical inventions (*Baco*, p. XCI).

the "light of order" when speaking of order as means for clearness in his methods in *Novum Organum* (Wallace 1973, p. 248). The motif of the importance of a secure method in searching for new truths and the motif of method serving as a "beacon of light" in this search is also found in Descartes' works. In *Rules for the Direction of the Mind* (1628), for example, Descartes claims in the fifth rule that method consists entirely in order and disposition of the objects we inquire about (*CSM*, Vol. 1, p. 379). In the tenth rule he mentions that he arrives at new truths via a secure method, not "blind inquiry" (*CSM*, Vol. 1, p. 403).[31] The importance and advantages of a secure method and order are central features in Leibniz's work as well:

> [...] the light of invention and the rigor of demonstration must be combined, and the elements of any discipline whatsoever must be written in such a way that the reader or disciple can always see the connection and, like a companion in the inventing, not so much follow the Teacher as walk along with him. (*Consilium de Encyclopedia nova conscribenda method inventoria* C 33, A VI, 4 A, N. 81, p. 341, in: de Risi 2007, p. 16)[32]

A similar statement appears in Maimon's work. He is not satisfied with the manner in which mathematics is taught, seeing that the student follows the teacher "blindly" and only imitates his actions. Like Leibniz, he wishes that the teacher instructs his students not only on the products of invention, but also on the ways and methods of invention. In his instruction, the teacher must aspire to have his students follow him as inventors and not as imitators. The inventor of a proposition or a solution to a problem first finds the proof or solution and only then arranges it as it appears to the student. Yet, the teacher presents the problem and its solution to the student and only then presents the proof (*Das Genie*, p. 374).[33] To use Leibniz's words, the teacher needs to teach the student

---

31 Method and order are very important features in Descartes' philosophy and mathematics. As mentioned by Grosholz, Descartes views all human knowledge as organized in such an order that one method is shared by all the sciences. She also mentions Vuillemin's remark that the model of Descartes' method is his theory of proportions: "the step-by-step constructions of his analysis are like the construction of unknown terms in a proportion. Just as the theory of proportions can be applied to any kind of mathematical object, so Cartesian method applies uniformly to all subject-matters" (Grosholz 2005, p. 6–7).
32 In *Two studies in the Logical Calculus*, Leibniz writes: "[...] this is the advantage of our method – we can judge at once, through numbers, whether proposed propositions are proved, and so we accomplish, solely with the guidance of characters and the use of a definite method which is truly analytic, what others have scarcely achieved with the greatest mental effort and by accident" (1969a, p. 236).
33 "The student can thus *understand* the proposition or the problem along with its solution, and *understand* the proof. But he will never *invent* the proof in this manner. Indeed, he cannot

how to find the pathways to solutions and proofs rather than follow the teacher blindly. That is the difference between an inventor of new methods and the imitator who can only repeat the work of others.[34]

In *Giv'at Hamore* (1791), Maimon praises the student who not only learns the words of the wise but also invents new truths. He emphasizes that invention cannot occur without order and method (p. 19).[35] One not only needs to use a method, but also to understand it. Otherwise, one is nothing but an imitator and cannot invent new truths on one's own. Maimon refers to this in *On Symbolic Cognition*,[36] when speaking of the difference between a "true philosopher" and a "philosophical calculator" (or "philosophical machine").[37] He distinguishes between a philosopher whose inventions are based on the inventions of others and one who only imitates and repeats the words of others without having insight into those "formulas" (*Tr.*, p. 282). Maimon presents three criteria for differentiating between the true philosopher and the philosophical calculator or, in other words, between the methodical inventor and the imitator. The first criterion is to be able not only to have a method, but also to understand it and the connections between its principles (*Tr.*, p. 282). Maimon does not use the word *method*, but *formulae*. Most likely he does so since he compares the philosopher to a

---

even understand how the teacher or the first inventor might have found it. He regards the inventor as a *wizard,* who conjures in front of him a yet unknown truth, as if it were a *new natural phenomenon*, without being capable to understand how this happens. Hence someone can know the entire work of Euclid, Archimedes, and God knows what else, without being able to invent the least of it himself; and even if he does, then this is not by a secure method but by chance" (*Das Genie*, p. 374).

34 A similar idea also appears in Bacon's work. Bacon describes seven kinds of methods (and mentions five more), one of them being the "Magistral" method which imparts knowledge to "the crowd of learners" for the purpose of using knowledge. Another method, the "Initiative", is taught in the way it was discovered to the "sons of science" for the purpose of communicating "the roots of his knowledge, not the fruits thereof" (Bacon, *De Augmentis*. VI.2; Works IV, 449; in: Wallace 1973, p. 244). In other words, the "Magistral" method teaches the products of other inventors, while the "Initiative" teaches one how to be a methodical inventor.

35 Freudenthal & Klein-Braslavy point to the significance of invention and refer the reader to Maimon's *Giv'at Hamore* (1791) where he declares that one needs to pursue the ways the inventors arrived at knowledge and not the truths themselves. By doing so, it is as if one becomes an inventor himself (Freudenthal & Klein-Braslavy 2003, p. 584; *Giv'at Hamore*, p. 5).

36 *On Symbolic Cognition and Philosophical Language* was published as a part of Maimon's first book, *Essay on Transcendental Philosophy* (1790).

37 In *Introduction to Logic*, published in 1800 (ten years after *On Symbolic Cognition*), Kant writes about the true philosopher in a manner that resonates with Maimon: "He then that desires to become, properly speaking, a philosopher, must exercise himself in making a free use of his reason, not a mere imitative and, so to speak, mechanical use" (Kant 1963, p. 183).

person solving algebraic problems using formulas. He quotes Kästner, who speaks of people who approach algebra without sufficient knowledge of geometry and "learn in some fashion to manipulate calculations with letters, but do not arrive at analysis itself, which is the guide of calculations" (*Tr.*, p. 280–281). Maimon comments that Kästner's remark is even more important for philosophy than it is for mathematics, since in mathematics one can follow a formula mechanically without understanding it and still arrive at a true result. In philosophy, however, following a formula mechanically can lead to no results, or even worse, false results. In philosophy, "the usefulness of the calculus depends on the correctness of the principles it starts out from" (*Tr.*, p. 282).

The second criterion is the manner in which one presents the work of others: the philosophical calculator uses the same expressions and order of presentation found in the system, whereas the true philosopher presents the work as if it were his own, thus making the work of others the basis for his own work (*Tr.*, p. 282–283). Maimon himself sets an example of how to be a "true philosopher" and not a "philosophical calculator." In his introduction to *Essay on Transcendental Philosophy*, he writes:

> The great Kant supplies a *complete idea* of transcendental philosophy (although not the whole science itself) in his immortal work *The Critique of Pure Reason*. My aim in this enquiry is to bring out *the most important truths* of this science. And I am following in the footsteps of the aforementioned sharp-witted philosopher; but (as the unbiased reader will remark) I am not copying him. I try, as much as it is in my power, to explain him, although from time to time I also make some comments [*Anmerkungen*] on him. (*Tr.*, p. 8–9)

Maimon's work is the work of a "true philosopher" since he searches for the rules used by Kant and develops his own philosophy in accordance. He does not follow Kant but, in Leibniz's words, he "walks along with him."[38]

The third criterion is often applied by Maimon in his works as well: the true philosopher, who knows that one must not only cite the method, but also un-

---

[38] In his autobiography, Maimon describes how he himself employs such a strategy: "I had now resolved to study Kant's *Kritik of Pure Reason*, of which I had often heard but which I had never seen yet. The method in which I studied this work, was quite peculiar. On the first perusal I obtained a vague idea of each section. This I endeavoured afterwards to make distinct by my own reflection, and thus to penetrate into the author's meaning. Such is properly the process which is called *thinking oneself into a system*. But as I already mastered in this way the systems of Spinoza, Hume and Leibniz, I was naturally led to think of a coalition-system. This in fact I found, and I put in gradually in writing in the form of explanatory observations on the *Kritik of Pure Reason*, just as this system unfolded itself to my mind. Such was the origin of my *Transcendental Philosophy*" (*Autobiography*, p. 279–280).

derstand it, should be able to follow his explanations with examples. Maimon claims that it is best to present examples from mathematics (*Tr.*, p. 283). Mathematics is an *a priori* science which determines both form and matter (since it constructs its objects), unlike philosophy, an *a priori* science that determines only form and not matter (since the objects are given), and also unlike the *a posteriori* natural sciences (*Tr.*, p. 2–3). Therefore, mathematics is the best science from which to present examples.[39]

Mathematics is the best of all sciences to serve as the platform for "perfecting the inventive faculty" of the methodical inventor. The methodical inventor does not simply "follow blindly" the practices of mathematics, but searches for the rules and methods according to which one can arrive at unknown truths. The next chapters examine how mathematics fills this role and how philosophy is enhanced by such methods.

1.4 The Objective Reality of the Genius and the Methodical Inventor

The genius and the methodical inventor differ in their ability to prove their objective realities. Maimon claims that it is difficult to prove the objective reality of a genius "and even more [difficult] to adjust its *application to particular cases*" (*Das Genie*, p. 368). Maimon also claims that the genius cannot even convince anyone that he possesses genius (or that he is a genius), let alone convince us of his genius by his inventions, which could also have been works of chance. For this reason, Maimon suggests "to leave [genius] in the hands of its lucky owner, and continue our investigation on the methods of invention" (*Ueber den Gebrauch*, p. 15–16).[40] Our investment in studying methodical inventions is worthwhile since it can be improved:

---

[39] The role of mathematics is very important in Maimon's philosophy in general, and in his theory of invention in particular: "Enough has already been said in praise of *mathematics* in general: namely that it perfects the cognitive faculty more than any other science by its evidence, order and certainty, rigor in proofs, etc. But how is this achieved? '*through practice*,' is the [common] answer. But this would mean: in an *obscure way*, as through practice we at last acquire a *skill* in a handcraft. Yet I would like to show that the adequate organized study of mathematics for the purpose of *perfecting the inventive faculty*, leads to this end step by step from the very beginning and not only through practice (such that we can account for every step taken)" (*Das Genie*, p. 372–373).

[40] Maimon also writes: "For that reason we want for now to gladly leave the undefined word *genius*, to the acrobats, French cooks and those fine spirits and artists that will be content with it, and look for our salvation elsewhere" (*Das Genie*, p. 365). Fichte refers to this passage

> Everybody speaks of genius; nobody of methodical inventions: yet still, we take more interest in the latter than in the former. The genius is a gift of nature, and we cannot contribute to its improvement; methodical invention, however, if it is *possible* as such, can certainly always be improved. (*Das Genie*, p. 371)

Moreover, Maimon states that if we wish to prove the objective reality of the genius in the sciences, we would have to go through the genius' works methodically, just as we would have done in the case of the methodical inventor. The difference is that the genius realizes his methods in an obscure manner, while the methodical inventor distinctly presents his methods (*Ueber den Gebrauch*, p. 14).

The objective reality of the methodical inventor lies at the center of Maimon's plan: "We must examine the reality of this concept [methodical inventor] in *actual inventions*, and abstract from it *methods of invention*" (*Das Genie*, p. 371). The actual inventions that are to be examined should not be taken from philosophy since it "produces merely *formal* inventions, but not *real* ones" (*Das Genie*, p. 372). Natural sciences make only discoveries and not inventions. Only mathematics, in which real inventions occur, is eligible for this purpose (*Das Genie*, p. 372). The following chapters demonstrate how to fulfill Maimon's plan of abstracting methods of invention from actual mathematical inventions.

---

in the second edition of *Concerning the Concept of the Wissenschaftslehre, or, of so-called "Philosophy"* (1798; first published in 1794). Fichte refers to Maimon's objection without mentioning Maimon's name: "I am not quite sure how and why, but an otherwise admirable philosophical author has become a bit agitated over the innocent assertion contained in the foregoing note. 'One would,' he says, 'prefer to leave the empty word "genius" to tightrope walkers, French cooks, "beautiful souls," artists, and others. For sound sciences it would be better to advance a theory of discovery.' One should indeed advance such a theory, which will certainly happen as soon as science has reached the point where it is possible to discover such a theory. But where is the contradiction between such a project and the assertion made above? And how will we discover such a theory of discovery? By means, perhaps, of a theory of discovery of a theory of discovery? And this?" (I.73, p. 128). It seems that Fichte is much more influenced by the *Geniezeit* than by the 17th century approach of the "light of reason." Whereas Maimon disposes of the study of genius as something that cannot contribute to his theory of invention, Fichte's approach claims that at the beginning of formulating new knowledge one does not act according to rules, but only blindly (p. 127); the rules only follow what has been formulated, they are not what leads to the formation of knowledge. This approach is very different from Maimon's, who argues that the genius acts according to a rule, even when he is not aware of it. As for Fichte's criticism on theory of discovery, we will see in the next chapters what possible response Maimon could have given.

Chapter 2: An Art of Finding Arguments

This chapter is dedicated to three features of Maimon's theory of invention. The first regards a famous topic in the *ars inveniendi* tradition: the question of invention versus discovery. Maimon's approach to this debate is that despite their differences, both discovery and invention can be considered under the general term of *invention*. In the second part of the chapter, I argue that the best general description of Maimon's work on invention is as an art of finding arguments: finding propositions that can serve as premises and conclusions in arguments. This general definition of invention as creating syllogism begins with given propositions. Therefore, the last section is dedicated to the notion of the *given* within the framework of Maimon's theory of invention.

2.1 Invention and Discovery

Invention is a general term that includes two processes: invention and discovery. Such a description of the relationship between the terms of invention and discovery was popular in Maimon's time.[41] Maimon defines invention in the broader sense as "to bring out *unknown truths out of known* truths following *secure methods*" (*Das Genie*, p. 363; *Ueber den Gebrauch*, p. 10).[42] This definition emphasizes the notion that we arrive at new knowledge by using methods and not by mere chance.[43] This stands in contrast to the central role played by chance in discovery as described by d'Alembert in *Encyclopédie* (d'Alembert, "Découverte", in: *Encyclopédie*, Vol. 4, p. 705).[44] In the narrow sense, Maimon bases the difference

---

41 For example, it appears in Diderot and d'Alembert's *Encyclopédie* (Jacourt, "Invention", in: *Encyclopédie*, Vol. 8, p. 848).
42 This definition is not original, but a very common description of invention by philosophers. For instance, this terminology was used by Leibniz when he claimed that there are two kinds of truths that need to be established: vague and imperfectly known truths and unknown truths. We should employ the method of certitude on known truths and we should employ the art of invention on the unknown truths (Leibniz, 1965a, p. 183).
43 Chance and methodic work can be jointly used for inventive purposes. As mentioned by Hoyningen-Huene, some contemporary scientists attempt to invoke chance by using systematic methods in their search for new discoveries. For instance, by using a system in different experimental conditions so that new properties are revealed (Hoyningen-Huene, 2008, p. 177).
44 D'Alembert mentions how geometricians came up with new theorems by chance while trying to solve the problem of the quadrature of the circle (d'Alembert, "Découverte", in: *Encyclopédie*, Vol. 4, p. 705). He presents three kinds of discoveries. The first is finding entirely new ideas. Arithmetic is an example of such a discovery: the invention of the digits 1–9 and the introduc-

between invention and discovery on whether the new knowledge that arises is an object or an attribute: to invent [*erfinden*] is to present *a priori* a new object, whereas to discover (or to find) [*finden*] is to ascribe *a priori* a new attribute to an already given object (*Das Genie*, p. 363; *Ueber den Gebrauch*, p. 10). Maimon remarks that, for all purposes, in his discussion he uses the broader definition that encompasses both invention and discovery (*Das Genie*, p. 363–364; *Ueber den Gebrauch*. p. 10). Examples of Maimon's distinction between invention and discovery can be found in his work, *A Short Exposition of Mathematical Inventions* (*Kurze Darstellung Mathematischer Erfindugen;* 1793). Maimon uses the verb *to invent* [*erfinden*] to describe both new methods and new objects: new methods such as the method of tangents, the method of indivisibles and solutions to problems e. g. Del Ferro's solution of the cubic equation; and new objects such as the caustic curve, logarithms and new machines e. g. Newton's reflecting telescope.[45] He uses the verb *to discover* [*entdecken*] to describe natural laws and phenomena, new properties of given objects and new (already existing but unknown to that point) natural objects: natural laws such as the laws of impact; natural phenomena such as the progression of light rays; new properties of given objects such as finding new properties of crystalline; and new natural ob-

---

tion of zero, along with the manner in which these digits are positioned (to determine the value of the number) enabled a new and better way to calculate numbers. According to d'Alembert, this counts as an act of genius. The second is joining a new idea to a known one. Algebra, the science of representing all possible quantities by general characters, is an example of such a discovery. The third is uniting two known ideas. For instance, we can apply algebra to geometry in order to discover new truths, by presenting a curve using an equation with two variables. These three methods are also referred to as "the fruit of genius" (*Encyclopédie*, Vol. 4, p. 705–706). As discussed in the previous chapter, Maimon's notion of genius is one whose actions are guided by rules (even if the genius is not aware of these rules) and this notion stands in line with Maimon's notion of invention and discovery as not governed by chance.

D'Alembert argues that discovery and invention differ in their importance. Discoveries are more useful, curious and difficult, and inventions are less important (*Encyclopédie*, Vol. 4, p. 705). There is no such distinction in the level of importance between the two in Maimon's texts.

45 For instance, he uses the verb *to invent* in the following cases: "Eine der wichtigsten Erfindungen des *Descartes* ist die Methode der *Tangenten*" (*Baco*, p. 288); "*Cavaleri* erfand die Methode der Indivisibeln" (*Baco*, p. 282); "Tschirnhausen erfand die Caustische Linie" (*Baco*, p. 299); "*Scipio Ferreo* erfand nach *Cardans* Bericht die Auflösung dieses Falles der cubischen Gleichung $x^3 + px = q$" (*Baco*, p. 270); "Tschirnhausen erfand die Caustische Linie" (*Baco*, p. 299); "Der Schottische Baron *von Nepper* erfand die *Logarithmen*" (*Baco*, p. 279); "*Newton* [...] erfand auch das Spiegelteleskop" (*Baco*, p. 305).

jects such as Saturn's ring, found by Huygens.⁴⁶ Accordingly, Maimon states that in the natural sciences objects are discovered, and that it is only in mathematics that new real objects can be invented (*Das Genie*, p. 372).

Maimon maintains that inventing and discovering presuppose conceiving [*begreifen*]. Conceiving is defined as "to derive the *possibility* of a *more determined* [object] from the already given *possibility* of a *less determined* (more general) object; or [to derive] the *necessity* of a *more determined a priori* proposition from the *necessity* of an already given *less determined* [proposition]" (*Ueber den Gebrauch*, p. 5). He presents the example of an equilateral triangle: its negative possibility is perceived *a priori* since its concept does not contain contradiction and its positive possibility (possibility of construction) is also perceived *a priori*, since the objects from which it is constructed (two circles whose radii serve as the triangle's sides) are given in intuition (*Ueber den Gebrauch*, p. 6).⁴⁷ This definition of conceiving leads Maimon to define the term in a more general manner as "to have a *universal concept* of it" (*Ueber den Gebrauch*, p. 6). Since the universal is included in the particular and what applies to the universal must apply to the particular, then in cases where the universal is already given in cognition, the particular is perceived *a priori* (*Ueber den Gebrauch*, p. 6). However, in his works on invention, he no longer uses this definition of conceiving as that which encompasses invention and discovery. The general definition that repeatedly appears under the general term of *invention* is the definition based on syllogisms.

In order to invent, writes Maimon, we should find propositions that can serve as premises or conclusions (*Ueber den Gebrauch*, p. 12–13). Using syllogisms, one begins with premises or a conclusion and by using middle terms arrives at the new truth.⁴⁸ Even though the syllogism serves as the basis of invention,

---

46 The verb *to discover* is employed in the following cases: "*Wallis, Wren*, und *Huyghens*, endeckten die Gesetze des Stosses der Körper" (*Baco*, p. 300); "Die *Platoniker* entdeckten die geradlinige Fortschreitung des Lichts" (*Baco*, p. 278); "Keplern [...] Er entdeckte den wahren Gebrauch der Crystallinse" (*Baco*, p. 291); "Huyghens entdeckte Saturnus Ring und den vierten Saturnustrabanten" (*Baco*, p. 303).
47 For Maimon's definitions of positive and negative possibilities, see *Tr.* p. 100–101.
48 Maimon uses the German word *Glied* when referring to middle term (e. g. *Ueber den Gebrauch*, p. 13). But the term *Glied* can also be used for a shared term that refers to a physical phenomenon. Lambert offers an example of such a case: Otto Von Guericke discovered the air pump by using the analogy between water and air. It is an analogy between a term [*Glied*] in the *given* and a term in the *sought* (Lambert 1764, § 485, p. 311). Von Guericke's invention appears a few times in Lambert's *Neues Organon*. It is also mentioned by Maimon, when commenting that von Guericke's air pump is a machine which was invented following the laws of elasticity of air (*Ueber den Gebrauch*, p. 7).

Maimon finds it to be an insufficient tool since it does not provide us with the means to find the sought premises (*Das Genie*, p. 377; *Ueber den Gebrauch*, p. 22–23). Nevertheless, this insufficiency of the use of syllogisms to provide us with the sought premises does not contradict the fact that the most basic and recurring concept of invention that appears in his articles on invention is based on syllogisms.[49] Furthermore, the methods presented by Maimon are meant to serve as tools to help us arrive at new propositions. So, for instance, in the method of analysis of the object we would use construction to arrive at a new premise that is the conclusion of the logical analysis being conducted. When discussing invention in general, Maimon refers to it as forming syllogisms.[50] It is only when presenting his theory in a more detailed manner that he introduces the different notions of invention (discussed in chapter 3).

---

49 In arguing for the insufficiency of the use of syllogisms for invention, Maimon may have been influenced by Wolff. According to Corr, Wolff stated that logic alone is insufficient for invention and that there is a need for other rules (Corr 1972, p. 330–331). However, even though Wolff at first regarded syllogism as a method that does not advance knowledge but only restates what is known, he later changed his mind and claimed that syllogisms can in fact serve as a rigorous tool to lead to new and certain knowledge (Corr 1972, p. 327).

50 In Maimon's article *The Genius and the Methodical Inventor*, one of the mentioned differences between the genius and the methodical inventor is that the latter is given the premises that lead to the unknown conclusion, or is given the conclusion, whereas the genius is not given any proposition but has to look for the propositions in his cognition (*Das Genie*, p. 366–367; *Ueber den Gebrauch*, p. 12–13). For more examples of Maimon's use of this general concept of invention as syllogism, see *Ueber den Gebrauch*, p. 2; p. 14–15; p. 22.

The question of the relation between syllogism and invention is vast and beyond the scope of this work. Therefore, it will not be addressed here at length. I would like only to mention that for Leibniz, who was an influential figure for Maimon, the art of syllogism is an important and powerful tool for invention: "Je tiens que l'invention de la forme des syllogismes est une des plus belles de l'esprit humain, & même des plus considerables. C'est une espece de Mathematique universelle, dont l'importance n'est pas assez connue" (Leibniz 1765b, Livre IV, Ch. XVII, § 4, p. 446). The inclusion of the art of syllogism as a kind of *mathesis universalis* means its inclusion as a kind of *ars inveniendi*. In Leibniz's time, the term *mathesis universalis* was considered as the model for all *ars inveniendi*. It was regarded as a science of quantity and of quality, magnitude and similitude, determined quantities as well as undetermined quantities (Rabouin 2002, p. 672–673). One of the main features of *mathesis universalis* is that it is based on relations (between magnitudes as well as between quantities): for instance, Marinos believed that Euclid's *Data* can serve as universal mathematics because it is a science of relations and proportions (Rabouin 2005, p. 255–256) and algebra was regarded by mathematician such as Descartes and Viète as universal mathematics that can serve also as an art of invention (Rabouin 2016, p. 265–266).

2.2 An Art of Finding Arguments

Maimon's *ars inveniendi* descends from a long tradition of theories of invention and discovery.[51] Maimon's theory of invention was influenced by many works on invention, such as those of Bacon, Descartes, Leibniz and Lambert. But perhaps the greatest influence was the art of finding arguments, originating in Aristotle's theory of Topics. *Ars inveniendi* as an art of finding arguments offers formulas for finding arguments and deals with the transition from an argument to a conclusion (Kienpointer 1997, p. 225).[52] Aristotle's *Topics* is dedicated to finding "a method by which we shall be able to syllogize about every problem that has been put forward from things approved" (Aristotle, *Topics*, 100a18 – 21; in: Alexander of Aphrodisias 2001, 1, p. 3).[53] Thus the first and most important influence on Maimon's work is the definition of invention (in general) as finding propositions to form syllogisms.[54] Moreover, when Maimon mentions that in

---

[51] Theories of both invention and discovery are known by the name *ars inveniendi*. This term is derived from Cicero's *inventio* (Kienpointer 1997, p. 225).
[52] Some claim that it is the first art of finding arguments (Kienpointer 1997, p. 226), whereas others claim that theories of topics existed before Aristotle's time (Green-Pedersen 1984, p. 20). Aristotle does not define or explain what a topic is, and he himself states that he inherited the idea from earlier thinkers (Green-Pedersen 1984, p. 20). So most likely his theory of topics is merely the first one that we encountered rather than the first to be conceived. Topics theory is also described as "a technique for testing arguments or commonly held opinions" (Green-Pedersen, 1987, p. 408) and as a technique for guaranteeing the transition from argument to conclusion (Kienpointer 1997, p. 226). It is suggested by Rubinelli that *topos* was used as a technical military term for "strategies for gaining the upper hand" in pre-Aristotelian rhetoric usage. She claims that it explains why Aristotle did not present a definition of the term, but treated it as a familiar one (Rubinelli 2006, p. 269).
[53] The name of this theory is derived from the Greek word *topos*, which means 'place' (Green-Pedersen 1987, p. 407). *Tópoi* are described as "'places' where arguments can be found" (Kienpointer 1997, p. 226) and as "'the point of departure' for an argument" (Green-Pedersen 1987, p. 407). They are also referred to as "sources" of premises and arguments (Spranzi 2011, p. 34). In Latin, *topos* is called *locus*, and both terms refer to the theory of topics. Scholars usually use the term *topos* when referring to Aristotle and the term *locus* when referring to Cicero's theory of topics, yet, as Green-Pedersen mentions, one can find the term *locus* used by Aristotle as well (Green-Pedersen 1987, p. 408). In his theory of topics, Cicero uses the term *inventio*, which is the translation of the Greek word *heuresis* for locating arguments (Bernard 2003b, p. 406) and the term *dispositio* for building arguments (Zompetti 2006, p. 20).
[54] *Ars inveniendi*, often translated as "art of finding" (e.g. Kienpointer 1997, p. 225; Sgarbi 2016, p. 43), refers not only to the art of finding arguments, but also to finding objects, including objects given *a posteriori*. Maimon's *art of finding* in the broader sense includes finding objects *a posteriori* such as the moons of Jupiter (*Baco*, p. 293). However, this kind of art of finding is not at the center of his work on invention – the art of finding arguments is.

order to find what is sought, something must be given, he writes that it is "the ontological law of the ancients" (*Ueber den Gebrauch*, p. 8). This assertion, originating before Aristotle's time, appears at the beginning of *Topics*.[55] It is unclear whether Maimon read Aristotle's *Topica*, but the text was likely available to him.[56] Regardless of whether there was a direct influence of Aristotle's work on Maimon's conception of invention, Maimon's notions and methods of invention are closely linked to such tradition of invention as an art of finding arguments.

Another influence of the art of finding arguments on Maimon's work is found in his methods of invention which he chose to include in his articles on the subject: all of these methods are meant to be used on Euclidean geometry and, as Maimon argues, involve finding either premises, middle terms or conclusions. Even though Maimon's work throughout the years included other kinds of methods of invention, the only ones he chose to include in the works dedicated to invention are the ones that can be applied to Euclidean geometry. Furthermore, the seven kinds of analysis presented by him refer explicitly to both problems and propositions. For instance, "analysis of a problem or a proposition, into its conditions" (*Das Genie*, p. 378; *Ueber den Gebrauch*. p. 24) and "analysis of the various ways according to which a problem can be solved or a proposition proven" (*Das Genie*, p. 380; *Ueber den Gebrauch*, p. 31). Referring to propositions and problems as two different types of knowledge is a feature of Greek mathematics.[57] In *Topica*, Aristotle claims that propositions and problems are equal in

---

[55] "Now a syllogism is an utterance in which, certain things having been posited [something different from these suppositions comes about by necessity through the supposition]" (Aristotle, *Topics*, 100a25–7; in: Alexander of Aphrodisias 2001, 1, p. 8).

[56] Aristotle's book *Topica* is a part of his *Organon*. In Maimon's translation and interpretation of Aristotle's *Categories* (*Die Kathegorien des Aristoteles*; 1794), he mentions that the translation was made according to Prof. Buhle's Latin translation of Aristotle's *Categories* (*KdA*, p. VII). In 1791, Theophilus Buhle published the first volume of his *Aristotelis Omnia Opera Graece* which includes a Latin translation of *Categories*. The third volume, published in 1792, includes the translation of *Topica*. Furthermore, Freudenthal & Klein-Braslavy mention that Maimon defines *accident* very similarly to Aristotle's definition in *Topica* (Freudenthal & Klein-Braslavy 2003, p. 590), which might suggest he either read this text or was indirectly exposed to its content. Lastly, the meaning of invention as an art of argumentation was known in Maimon's time. For instance, Jacourt refers to invention as the art of argumentation (*Encyclopédie*, Vol. 8, p. 849).

[57] Knorr mentions that later commentators of Greek geometry exceedingly emphasized the theoretical part over the problematic one. He explains that theorems overshadow problems as the result of being "blinded by the glare of Plato's ontology" (Knorr 1993, p. 360), where in fact, geometers such as Apollonius concentrated in their analysis on finding solutions to problems (Knorr 1993, p. 360). In *Collection*, Pappus mentions Euclid's porisms as a third type of knowledge, that is "neither theorems nor problems, but of a type occupying a sort of mean between them, so that

number.⁵⁸ It is very likely that Maimon's explicit mentioning of both propositions and problems in his methods of analysis was influenced by Aristotle either directly or via Greek geometry.⁵⁹

The presentation of methods that are aimed at finding propositions to be used in proofs and solutions affiliates Maimon's work to the notions of invention of Greek philosophers and mathematicians rather than to the works of more recent philosophers and mathematicians, such as Descartes and Leibniz. Unlike Descartes' introduction of the unit in his *Regulae* and the presentation of a new theory of proportions between magnitudes, and unlike Leibniz's *ars characteristica* and *ars combinatoria*, Maimon does not present new mathematical objects or new mathematical (or logical) systems.⁶⁰ He only presents methods that

---

their propositions can assume the form of theorems or problems [...] That the ancients best knew the distinction between these three things, is clear from their definitions. For they said that a theorem is what is offered for proof of what is offered, a problem what is proposed for construction of what is offered, a porism what is offered for the finding of what is offered" (Pappus, *Book 7 of the Collection* (13–14), from: Jones 1986, p. 94–96).

58 Every argument can be phrased as a problem and vice versa, since "every proposition and every problem indicates either a genus or a peculiarity or an accident" and since "The problem and the proposition differ in the way in which they are stated," so that by altering the statement we can change one by the other (Aristotle 1976, *Topica* Book I, IV 101b11–37).

59 According to Proclus, Euclid distinguished between problems and theorems based on whether a new object arises (problems) or a new property (theorems; Proclus 1992, p. 63). Following Maimon's definitions of invention and discovery we can assert that in problems we invent and in theorems we discover. Proclus explains that the difference between problems and theorems is based on whether the new predicate is possible or necessary: "When, therefore, we propose to inscribe an equilateral triangle in a circle, we call it a problem, for it is possible to inscribe a triangle that is not equilateral; or again to construct an equilateral triangle on a given line is a problem, for it is possible to construct one that is not equilateral. But when a man sets out to prove that the angles at the base of an isosceles triangle are equal, we should say he is proposing a theorem, for it is not possible that the angles at the base of an isosceles should not be equal. Thus if anyone were to set it up as a problem to inscribe a right angle in a semicircle, he would be regarded as being ignorant of geometry, for any angle inscribe in a semicircle is a right angle. In general, then, all cases in which the property is universal, that is, coextensive with the whole of the matter, must be called theorems; but whenever the character is not universal, that is, does not belong to the whole genus of the subject, then it must be called a problem." (Proclus 1992, p. 65). In chapter 3, I discuss Maimon's definitions of analysis and synthesis as based on necessity and possibility, as well as their relation to the notions of discovery and invention.

60 Leibniz's *ars inveniendi*, *ars characteristica* and *ars combinatoria* are all connected (Pasini 1997, p. 41). For instance, in his mathematical dictionary, Wolff dedicates an entry to *ars combinatoria characteristica* (*Vollständiges mathematisches Lexicon* 1734, p. 105). Leibniz's *ars inveniendi* is used in his *ars combinatoria* to develop a rational structure, and also to apply this structure to a contingent context (Van Peursen 1986). Ars *charcteristica* serves *ars inveniendi*

he extracts from actual invention, methods that are to be used on propositions and problems of Euclidean geometry. Maimon's choice to present seven kinds of analysis that are based on methods used in Euclid's *Elements* rather than on the newer sciences of analysis which he had knowledge of, such as algebra, can only be explained by his notion of invention as an art of finding arguments. The choice of this notion for invention is even more astounding when we consider that in the 17th and 18th centuries, the status of topics theories was usually much lower than that of invention theories (even though the latter originated in the first).[61]

Another feature of Maimon's work on invention, which might not be directly influenced by topics theory but is still worth mentioning, is his reluctance to present a general science, presenting instead only methods for mathematical problems.[62] In this, Maimon's theory of invention resembles Cicero's theory of *loci*, which is a theory of topics. Cicero's *loci* theory does not offer a general method of invention, but rather strategies for discovering material in particular cases (Leff 1996, p. 450).[63] Maimon provides methods for mathematics, and more specifically, only for certain kinds of mathematical problems – problems of Eucli-

---

by exposing hidden structures and properties (Knobloch 2010, p. 291). The main purpose of Leibniz's *ars characteristica* (also known as algebra or analysis) is to free ourselves from imagination by placing signs instead of the things (similar to numerals in arithmetics or notes in music) (Leibniz 1961a, p. 97). He was interested in the application of the *ars characteristica* to geometry, and this idea seems to have come to him after reading the first book of Euclid's *Elements* (Leibniz, GM., V, p. 183–211 in: Leibniz 1969b, editor's note, p. 248). This method is mentioned by Maimon as one by which we can arrive at new mathematical inventions by using general algebraic signs (*Erfindungsmethoden*, p. 139).

61 For instance, Bacon considers invention as one of the four logical arts, which is divided into the discovery of arts and arguments. The latter includes the topical method (Bacon 1859, Vol. 1, *An Analytical View of the De Augmentis Scientiarum*, p. lxxvii). Arnauld and Nicole's *Logic, or The Art of Thinking* includes a chapter with the title: "Topics, or the method of finding arguments. How useless this method is" (1996, Third Part, Ch. 17, p. 232). Leibniz considered topics only as tools for improving memory and order, i.e., as important tools for rhetoric but not for finding hidden necessary truths (Leibniz 1765a, p. 532). In the first edition of *Critique of Pure Reason*, Kant regarded topics only as an aid to memory, as a tool whose use grants "an appearance of thoroughness" (*CpR*, § A 268–9/B 324–5; in: Capozzi 2006, p. 145). However, with time, Kant came to acknowledge the importance of this tool that can be used for analyzing an object by presenting different points of view to treat it (Capozzi 2006, p. 144–146).

62 It should be noted that this reluctance is in relation to his theory of invention. As mentioned by Gueroult, Maimon believed that philosophy could serve as the science of all sciences, a general science that would enable the systematization of our knowledge (Gueroult 1929, p. 12).

63 The term *loci* serves in geometry to describe both extension and situation (or position). In Leibniz's *analysis situs*, for instance, a plane is "the locus of all points which, one by one, bear the same situation" (De Risi 2007, p. 215–216).

dean geometry.⁶⁴ He explicitly writes that he does not wish to present a complete theory of invention that encompasses all kinds of human knowledge (including fields such as ethics), as did Leibniz and Wolff, unsuccessfully.⁶⁵ The general science proposed by Leibniz, which he refers to as the art of invention, is meant to find the general truths from which the rest of the propositions follow (Leibniz 1765a, p. 529).⁶⁶ Familiar with Leibniz's work, Maimon refers to Leibniz's theory of invention as not fulfilling the promise it entails (*Ueber den Gebrauch.* p. 2). He believes that Leibniz's project is important but very difficult to execute (*Tr.*, p. 324) and chooses to present only a few rules of invention, all closely connected to actual inventions and examples of particular cases. Unlike Leibniz's sug-

---

64 The relations between mathematics and rhetoric (and more specifically, topics theories) are very strong, even lasting until today. Thus, for instance, rhetorical tools were used by Pappus when solving mathematical problems (Bernard 2003b, p. 406). In the 16th century, one of the aims of Peletier's algebra, which he named *art of thinking*, was "the production of valid arguments" (Cifoletti 2006, p. 370). Perhaps the most interesting example is Alexander Grothendieck's *topos* theory, which is a mathematical theory in the field of algebraic geometry invented in mid-20th century. Grothendieck's choice of the term *topos* to indicate a new mathematical object was directly influenced by the meaning of the Aristotelian *topos* as "place" (Illusie 2004, p. 1060). In addition, according to Netz, Greek mathematicians were very influenced by the culture and philosophy of the time, and especially by rhetoric. The influence was so strong that mathematicians regarded themselves as 'authors' rather than 'practitioners' (Netz 1999, p. 311).
65 Maimon writes: "[...] they searched for a *general* complete *theory of invention*, extending on the *whole extension of human cognition*. I, however, want to limit myself in this only to *mathematics* and *physics*. Also, I want no *complete theory*, but rather merely to abstract every now and then from these sciences those *methods* revealed to me and also to prove its *possibility* of the matter by the *fact* [*Factum*] itself. Presupposing the existence of *sufficient acquired knowledge*, everyone will be able to test whether he is able to *invent* something new by these *methods*" (*Ueber den Gebrauch*, p. 4).

Leibniz's works on *ars inveniendi* concentrate on mathematics, but that is not to say that he narrowed his interest to developing a theory of invention for mathematics alone. In *Projet d'un Art d'Inventer*, Leibniz expresses his wish to have a theory of invention for metaphysics, physics and ethics as well, with the same certainty as for the elements of mathematics (Leibniz 1961b, p. 175). In *Nouvelles ouvertures*, Leibniz writes on the two projects of inventory and general science and suggests that the latter must begin before the former (Leibniz 1961c, p. 228–229).

It is worth mentioning that in Carl Friedrich Flögels' work on invention, published in 1760, the Jewish Kabbalah was included as art of invention, under the title of "Characteristics". Among the methods for finding new things, such as Ziruphe and the use of Amuletes, he mentions Gematria – assigning of a numerical value to a word (§ 170, p. 163–164). Due to Maimon's rational notion of invention as forming syllogisms and the inventor as one who follows rules, it is quite understandable why this subject matter is not part of Maimon's theory of invention.
66 It seems that general science was of such importance for Leibniz that he even wrote that he valued mathematics only for the sake of finding in it "traces of a general art of invention" (Goethe et al. 2015, p. 13).

gestion to begin with general science and only then turn to the project of inventory of knowledge (Leibniz 1961c, p. 228–229), Maimon chooses to start with the inventory.[67] In *On Symbolic Cognition* (1790), he describes the philosophical dictionary he will publish one year later as "a collection of definitions" (*Tr.*, p. 332). It is based on simple concepts that can serve as a basis for a universal philosophical language, but this language will not be used for scientific discovery (*Tr.*, p. 330–332). The development of a general science is left to future mathematicians, whom Maimon hopes will follow his work. His wish is that through their work, "mathematics will thereby swing upwards to the rank of the *science of all sciences* by presenting general methods, theories and systems for the expansion of knowledge as such, and by establishing them for general use" (*Das Genie*, p. 383–384).

Maimon was familiar with different fields of mathematical knowledge. For instance, he wrote a textbook on algebra in Hebrew (*Ma'aseh Choshev*).[68] He received a formal mathematical education when he began studying in the gymnasium in Altona in 1783, where he spent two years. He studied mathematics under the director, Mr. Dusch, who lectured on Segner's *Mathematical Compendium* (*Autobiography*, p. 258–259)[69] and when he was a tutor, he taught Euler's *Algebra* (*Autobiography*, p. 274). His choice to establish his theory of invention on, and only on, examples from Euclidean geometry could have two origins. The first, being that ancient Greek geometry has stronger relations with rhetoric than other mathematical fields (which Maimon had relatively good knowledge of),

---

[67] We can assume that Maimon was acquainted with Leibniz's distinction between inventory of knowledge and general science from *Discours touchant la methode de la certitude et l'art d'inventer* (Leibniz 1765a, p. 531), since the text was explicitly mentioned by Maimon (*Ueber den Gebrauch:* 2).

Leibniz's distinction between ordering knowledge that is already known (*ars demonstrandi*) and unknown knowledge to be discovered (*ars inveniendi*) had already been formulated by Cicero (Van Peursen 1986, p. 183–184).

Such a work of inventory is Locke's *A New Method of a Common-Place Book* (1687). It is an index of *topoi* (*places*) or *loci* (*commonplaces*), which were used by philosophers from the Renaissance until the Enlightenment as a source of knowledge to be quoted and referred to (Walker 2001, p. 114).

[68] For example, in *Ma'aseh Choshev* Maimon solves polynomial equations with three variables and unknown powers (*Hesheq Shelomo*, [manuscript; 1778], p. 245). As mentioned by Freudenthal, it is unclear when *Ma'aseh Choshev* was written and we cannot assume that it was written at the same time as the other four works in *Hesheq Shelomo* (approximately in 1777; Freudenthal 2011, p. 114). However, it is almost certain that he acquired his algebraic knowledge before publishing his articles on invention.

[69] He attended the Altona Gymnasium Christianeum (Freudenthal 2004, p. 128).

such as algebra, analytical geometry or calculus. By defining invention in general as based on syllogisms, the choice of Greek geometry as the science on which to demonstrate the invention of new premises or conclusions seems obvious. The second is Maimon's wish to establish his theory of invention not only on extracting methods from given inventions but more specifically, on inventions given in actuality.

Maimon's appeal to inventions that are given in intuition is due to the fact that what is given as actual is proven as possible as well.[70] He extracted all of the methods of invention (that are presented in his works on invention) from actual inventions in order "to prove its *possibility* of the matter by the *fact* [*Factum*] itself" (*Ueber den Gebrauch*, p. 4; *Das Genie*, p. 371–372). By beginning with the actual, he attempts to avoid what he believes to be the problematic part of the works of some of his predecessors. Such is Leibniz's *ars inveniendi*, which had not provided an answer to the question: how is invention possible? (*Ueber den Gebrauch*, p. 2). Maimon's answer to this question is based on actuality and Greek geometry, and being based on intuition, the latter is an excellent choice to serve as the science from which to extract methods.

Actuality is defined by Maimon as being represented in time and space (*Tr.*, p. 248).[71] Perhaps the strongest evidence for the importance of actuality in Maimon's published works on invention is his exclusion of other methods of invention that are based on objects which are not given in actuality. Methods such as the method of fictions are declared by him as important in other works, but they are not mentioned in his works on invention. For example, in the essay *On the Progress of Philosophy* [*Ueber den Progressen der Philosophie*, 1793], Maimon refers to fictions [*Fikzionen, Erdichtungen*] as means of invention of the greatest importance in mathematics and philosophy (*Prg.*, p. 54–55).[72] Among the methods of invention based on fictions, one can find Cavaleri's method of indivisibles, differential calculus and the method of tangents (*Prg.*, p. 17–18). Fictions are defined by Maimon as that which is possible but not actual (*Tr.*, p. 103–104). Although fictions themselves are not real objects, the relations between fictions

---

70 The actuality of the sought is determined by the actuality of the given. Therefore, since the sought is actual it is also possible (*Ueber den Gebrauch*, p. 8; See also Schulz 1954, p. 284).
71 "According to the Leibnizian-Wolffian school, actuality is the complete possibility of a thing. But on my theory, the actuality of a thing is its representation in time and space" (*Tr.*, p. 248).
72 This essay was submitted by Maimon for the 1792 annual essay contest of the Royal Academy in Berlin, on the subject of the progress made in metaphysics since Leibniz and Wolff. It was also published as the first chapter of *Streifereien im Gebiete der Philosophie* (1793).

can be used to determine real objects (*Prg.*, p. 17).[73] Even though Maimon mentions that real objects can be generated from symbolic cognition (such as fictions), Maimon's theory and methods of invention are not dedicated to real objects but to objects given in actuality.

Another example of a method of invention, which is productive but does not appear as a part of Maimon's work on invention, is the use of symbolic cognition. For Maimon, symbolic cognition is of great importance since it allows us to discover new abstract concepts from known ones by the use of reason (*Tr.*, p. 265). However, in order to be useful, Maimon believes it should be grounded in intuition (*Tr.*, p. 265).[74] For the reason that intuitive knowledge has an advantage over symbolic knowledge since using the latter we can arrive at knowledge that does not correspond to real objects, such as imaginary numbers (*Tr.*, p. 410–412). He refers to symbolic philosophical language, which is not grounded in intuition, as a means of learning and teaching truths but not of discovering new truths (*Tr.*, p. 326).[75] The symbolic infinite, together with the real infinitely small and the intuitive infinitely small, are mathematical inventions that are not mentioned as part of the methods of invention in Maimon's works on the subject.[76] The exclusion of these methods is due to the inability to present

---

[73] The faculty of imagination presents fictions as real objects, while reason declares them to be mere fictions (*Logik*, p. 205–206).
[74] Buzaglo describes the relation between intuition and symbolic cognition as a ladder, where the first step is cognition based on intuition and the last is pure symbolic cognition (Buzaglo 2002, p. 56). That is not to say that intuition is unimportant: "Intuition is a ladder on which we climb up to the forms – a ladder which *we do not* kick away when we have reached the top. Intuition is always in the background, for we always have to go back to it in order to progress further up the ladder" (Buzaglo 2002, p. 62).
[75] "How does it help me, for example, to have the correct definition of a hypotenuse, namely that it is the side of a right-angled triangle that is opposite the right angle; I will still never extract the claim that the square of the hypotenuse is equal to the sum of the squares of the other sides from this definition without construction and certain tricks in drawing some adjacent lines (called *artificia heuristica*); and it is the same in other cases too" (*Tr.*, p. 326).
[76] The symbolic infinitely small is defined by Maimon as "a state that a quantum approaches ever closer to, but that it could never reach without ceasing to be what it is, so we can view it as in this state merely symbolically" (*Tr.*, p. 351). He mentions the cosine of a right angle as an example: if two parallel lines make an angle with one another, then either the angle itself is infinitely small and ceases to be an angle, or the cosine is infinitely small and ceases to be the cosine. In both cases, these are only limit concepts and not quanta that are given in intuition (*Tr.*, p. 351–352).
The real infinitely small is defined as what "can itself be thought as an object (and not merely as the predicate of an intuition) despite the fact that it is itself a mere form that cannot be constructed as an object, i.e. presented in intuition" (*Tr.*, p. 353). An example is the absolute unit in pure arithmetic (*Tr.*, p. 353). The intuitive infinitely small is defined as quantum that

these objects in intuition and a result of the importance that Maimon grants actuality in his works on invention.[77]

The appeal of Euclidean geometry to Maimon, rather than other mathematical fields such as algebra and calculus (which are based on objects generated as symbolic cognition), might also be a result of his notion of invention as an art of finding arguments. The choice of Euclidean geometry as the exemplar science on which to conduct his methods of invention, together with the choice of defining invention as an art of finding arguments, is a choice to concentrate on a field in which intuition and understanding meet, or at least approach each other. In Euclidean geometry, we infer from propositions which are grounded in intuition. Perhaps the best articulation of this approach is when Maimon describes how in order to find the sought term of the concept of the object, we must perform an analysis of the object (*Ueber den Gebrauch*, p. 13–14). Moreover, by applying methods of an art of finding propositions to Euclidean geometry he might have hoped to demonstrate that geometry (or some of its propositions, at least) can be reduced to logic. In one of his methods, Maimon attempts to do exactly that: to prove a problematic proposition as true and necessary by arriving at an identity judgment at the end of a geometrical demonstration (although unsuccessfully; see Section 4.4).

Maimon's choice of Euclidean geometry as the science standing at the center of his theory of invention raises the question of whether he considered geometry to be a science of figures or a science of space.[78] On the one hand, Euclidean geometry is what De Risi refers to as "a geometry without space" (De Risi 2015,

---

ceases to be determined and is only determinable. It does not cease to be quantum (as in the case of symbolic infinitely small), but it is not a determined quantum (*Tr.*, p. 351). For instance, the differential of a magnitude: the ratio between the differentials dx and dy is compared with the ratio between the given magnitudes a and b ($dx:dy = a:b$). Since nothing cannot be proportional to nothing, the differential is a quantum, but only a determinable quantum. Therefore, explains Maimon, the differential dx does not mean that x is abstracted of any magnitude, but that we can take x to be as small or as large as we want. That is, dx is undetermined, but does not cease to be a magnitude (*Tr.*, p. 352).

77 The exclusion of fictions from Maimon's articles on invention shines in even brighter light when we consider that, as Atlas remarks, Maimon sometimes used the term "fiction" to mean "method" (Atlas 1969, p. 366). This exclusion is due to Maimon's preference to present methods based on Euclidean geometry, i.e. on objects given in actuality.

78 According to de Risi, the first articulation of geometry as a science of space was Leibniz's *analysis situs* (De Risi 2015, p. 1). It dealt with the study of properties of space such as uniformity, homogeneity, continuity and more, and is referred to by De Risi as "an abstract study of the structural properties of a system of relations" and as "a science of position and place" (De Risi 2015, p. 10).

p. 2). It deals with magnitudes and with individual figures as unrelated rather than with space and relational structures (De Risi 2015, p. 2–3). On the other hand, Maimon explicitly writes in his first published book that the object of pure geometry is space and he defines it as relation.[79] Moreover, according to his principle of determinability, geometrical objects are generated by determination so that the triangle, for example, is a determination of space: "The concepts of the objects of *mathematics* (produced [*hervorgebrachten*] by thought according to the principle of determinability) belong to *synthetic knowledge*. For instance, the concept of a triangle originates from the synthetic judgment: "*space can be enclosed by three lines*" (*Logik*, p. 123).[80] If we consider Maimon's conception of geometry according to his work on invention alone, then the use of Euclidean geometry as the science on which he develops his methods leads us to determine that *de facto*, in the cadre of his work on invention, he treats geometry as a science of figures rather than a science of space.[81] As will be shown in chapter 4, the practice of his methods is on individual figures and magnitudes and not on space as a relation structure. Even though Maimon himself defines space as relation (*Tr.*, p. 18),[82] this definition has no use in, or influence on his practice of geometry, at least as far as it enters into the application of his methods of inven-

---

[79] "The object of pure arithmetic is number, whose form is pure time as a concept; on the other hand, the object of pure geometry is pure space, not as concept, but as intuition. In the differential calculus, space is considered as a concept abstracted from all quantity, but nevertheless considered [as] determined through different kinds of quality in its intuition" (*Tr.*, p. 22–23).
[80] Freudenthal mentions Maimon's example of the conic sections as an example of construction of geometrical objects from a more complex one: we can construct a circle, an ellipse, a parabola, a hyperbola, a straight line and a point from a circular cone (Freudenthal 2010, p. 100).
[81] Were Maimon to consider geometry as a science of space (in the cadre of his articles presenting a theory of invention), he would not have presented methods to be applied on Euclidean geometry and would not have outlined his theory as an art of finding propositions. But rather, he would have presented rules of determinability, such as the requirement that each predicate has only one subject (e. g. *Tr.*, p. 380), attempting to show how triangles are determinations of space. Or, he would have presented new mathematical concepts, definitions and notations to establish this new geometry, as did Leibniz when introducing the new notation *A.B* to mean the situation between *A* and *B*. This notation served him in his *analysis situs*, a science in which there is no situation of a single object but only relation between a set of objects (De Risi 2007, p. 133).
[82] Maimon defines space both as intuition and concept: "Space and time are as much concepts as intuitions, and the latter presuppose the former. The sensible representation of the difference between determined things is the being-apart of those things; the representation of the difference between things in general is being-apart in general, or space. As unity in the manifold, this space is therefore a concept. The representation of the relation of a sensible object to different sensible objects at the same time is space as intuition" (*Tr.*, p. 18).

tion. Although on many occasions Maimon emphasizes the importance of the principle of determinability in the creation of geometrical objects, he does not mention it at all in the works dedicated to invention. I have included this principle in my discussion of synthesis (Section 3.6) and of the method of logical analysis (Section 4.1.7.2), where I attempted to connect his work on invention with his work on syllogisms as based on the relation of determinability.

2.3 The Given

Invention is defined by Maimon in various ways. Common to all of these definitions, may it be as logical analysis or as synthesis, is that there is always something given at the beginning of the process.[83] The *given* is given propositions or conclusion (*Das Genie*, p. 366–367). A proposition can also be assumed as given (*Erfindungsmethoden*, p. 140). This corresponds to the form of hypothetical judgements in which, when something is given, so another thing is given.[84] In his methods, *given* can refer also to an object given in intuition. Maimon uses the rhetoric of both *given and sought* and *known and unknown*, two sets of terms that were traditionally used in theories of invention.[85] He quotes what he calls "the ontological law of the ancients": "*e nihilo nihil fit*, where nothing is given, nothing can be found" (*Ueber den Gebrauch*, p. 8) and refers to *given* and *sought* throughout his articles. He also describes invention as bringing unknown truths from known ones (*Ueber den Gebrauch*, p. 10). Beginning invention with something given is the most basic principle in his theory of invention. It is also the manner in which consciousness arises in our cognition (*Tr.*, p. 29; p. 350) and it characterizes any cognition as such. I argue that there are two types of *given* in Maimon's theory of invention, which I refer to as *known* [*bekannt*] and *given* [*gegeben*].

I suggest that in the broader sense, *given* is something that is presented to our cognition passively. This general notion includes both *given* and *known*:

---

[83] When relating to the possibility of a general theory of invention, Maimon writes that it is "based on the solution of this problem: *There should be given certain* [*sichere*] *methods, according to which we can determine out of the mass of the already acquired knowledge the given premises to any given conclusion, or to a conclusion in general*" (*Ueber den Gebrauch*, p. 2).
[84] This process cannot be conversed (*Logik*, p. 70).
[85] Historically, the transition from using the term *given* to using *known* was made by Arab mathematicians. It is this transition from 'the given' to 'the known' that later enabled the emergence of algebra, analytical geometry and infinitesimal analysis (Hintikka 2012, p. 65). Similarly, my use of the term *known* refers to knowledge that is not given in intuition.

given (in the narrow sense) means given in intuition, a representation that we perceive in the forms of intuition, space and time. This is the more common meaning of the two meanings of the term and it is the one Maimon uses more often in his philosophy. *Known* in the narrow sense means given as a proposition. A proposition can be given to the cognitive faculty in a passive way, without being thought actively and yet also not be intuitive. The need for the latter arises from Maimon's work on *given* in invention.

Maimon does not refer to this differentiation between *given* (in intuition) and *known*, but the need for it arises from his use of both terms. His common use of the term *given* in his philosophy is "given in intuition". In *Essay on Transcendental Philosophy*, for example, he defines *given* as follows: "'given' signifies only this: a representation that arises in us in an unknown [*unbekannt*] way" (*Tr.*, p. 203). The notion of *given* as what is given in intuition is also evident in commentaries to Maimon's philosophy. For instance, according to Atlas, "the 'given' must have extension" (Atlas 1964, p. 117). Kuntze states that we order the manifold of the given in time and space (Kuntze 1912, p. 74–75).[86] Consequently, I suggest defining given in the narrow sense as "given in intuition".

Maimon's most repeated notion of invention is the invention of new propositions from given propositions. In this case, the *given* appears in the sense of known, and not as given in intuition:

> Given the proposition, for example: *C* is *B* given as a conclusion, which I am supposed to prove from its *premises*. To this purpose I recall to my memory all propositions already known to me, reject those which share no term with this conclusion, for instance, *D* is *E*, *F* is *G*, etc., and keep the ones which share something with the former: for instance, *A* is *B*, *C* is *A*. These are now the *sought premises*. Here comes to help the faculty of imagination, so that indeed I do not need to pass within the critical examination all my already acquired knowledge, rather, since the *sought* propositions share *one* term with the *givens*, so I can get there without detours according to the law of association. (*Ueber den Gebrauch*, p. 12–13)

---

[86] See also: Atlas 1964, p. 67, p. 90 & p. 116.
  Sylvain states that *es gibt* has a logical meaning: to be given is to be constructible and presented as an object (in intuition). Using this definition of *given* he explains Maimon's inclination to be prudent when one uses the word *God*, to which we cannot present a given object (Sylvain 1986, p. 199–200).
  Two further examples of the use of *the given* as an object being given in intuition can be found in the discussions of Katzoff (1981, p. 88) and Herrera who refers to the *given* as given by sensibility (2010, p. 602–603).

2.3 The Given — 43

Therefore, we would broaden the definition of *given* to include not only sensible representations but also knowledge in the form of propositions. I suggest the definition of *known* in the narrow sense as given as a proposition.

In order to define the general notion of given, that includes both given in intuition and given as a proposition, we should turn to Maimon's description of the given as arising "outside us." It appears in our consciousness in a passive, rather than an active manner, i.e., not actively thought by our faculty of thought (*Tr.*, p. 202–203). Maimon defines *thinking* as an action, not a state:

> The understanding can only think objects as flowing [*fliessend*] (with the exception of the forms of judgement, which are not objects). The reason for this is that the business of the understanding is nothing but *thinking*, i.e. producing unity in the manifold, which means that it can only think an object by specifying [*angiebt*] the way it arises or the rule by which it arises [*die Regel oder die Art seiner Entstehung*]: this is the only way that the manifold of an object be brought under the unity of the rule, and consequently the understanding cannot think an object as having already arisen [*entstanden*] but only as arising [*entstehend*], i.e., as flowing [*fliessend*]. (*Tr.*, p. 32–33)

Thinking means thinking a rule for the way in which the concept arises. There are judgments that are not based on a rule but on intuition, where the presented connection between the subject and predicate is based on something given in intuition. Therefore, some judgments are not actively thought but presented passively to our cognition. Although they are presented passively and can also be based on intuition, they are not intuitions themselves. It is this status of some judgments as not intuitions and not thought concepts, that requires us to introduce the notion of *given* in the broader sense: something introduced passively to our cognition. This definition also accords with the notion of *given* in the narrow sense. When referring to the color red as something that is given to the cognitive faculty, Maimon states: "we say '*given*' [*gegeben*] because the cognitive faculty [*Erkenntnißvermögen*] cannot produce the colour out of itself in a way that it itself prescribes, but behaves merely passively in relation to the colour" (*Tr.*, p. 13). That is, we do not think red but only perceive it.

In given propositions, the ground for connecting a subject and a predicate may arise from intuition. Nevertheless, in this case the proposition itself is not an intuition given in time and space. A given proposition is something that was created actively by the understanding but is nevertheless given to us passively. It is not something that is passively given by the imagination, since what is given by the imagination is an object given in intuition and not a proposition.[87] It is thus

---

[87] For instance, Maimon mentions that it is the imagination that presents a concept in time and space (*Tr.*, p. 100; p. 135). He claims that the imagination is the ground for things that are outside

also not something produced by the productive imagination, which has not only a passive aspect, but also an active aspect of creating *a priori* synthesis: "it does not merely take objects all at once but orders them under one another and connects them" (*Tr.*, p. 20).[88]

The transition from *given in intuition* to *known* is due to the symbolic nature of a conclusion in syllogisms.[89] According to Maimon, arriving at conclusions is a symbolic process. Indeed, we think a plurality of cognition in unity of consciousness based on antecedent intuitive cognition, but in itself the process is not intuitive (*KdA*, p. 199).[90] In *Elements*, we prove propositions using intuition. However, once a proposition was proven, it is used as symbolic knowledge only. For instance, *Elements* I.4 is a conclusion of a proof conducted in intuition that is then used as given symbolic knowledge in another demonstration. Maimon describes how *Elements* I.4 is used as a premise in the proof of *Elements* I.5 (*Ueber den Gebrauch*, p. 30): To prove *Elements* I.5, we use the given object as it is given in intuition (we turn from the general proposition to a particular triangle *ABC* and to a diagram). When *Elements* I.4 is used as a premise in the proof of *Elements* I.5, we pay no regard to the objects that were given in the proof of *Elements* I.4 (triangles *ABC* and *DEF*), but only to the relations between them (the equality between two sides and the angle between them, in a set of two trian-

---

one another in time and space: "[...] the imagination, which is the ape of the understanding, represents the things *a* and *b* as external to one another in time and space" (*Tr.*, p. 133–134). Furthermore, he states that actuality is a synthesis made according to the laws of the imagination and not the laws of understanding (*Tr.*, p. 102).

Kant clearly defines the faculty of imagination as presenting objects in time and space: "*Imagination* is the faculty representing in intuition an object that is *not itself present*. Now since all our intuition is sensible, the imagination, owing to the subjective condition under which alone it can give to the concepts of understanding a corresponding intuition, belongs to *sensibility*" (*CpR*, § B151, p. 165).

88 Another faculty that is both active and passive is the faculty of fictions [*Erdichtungsvermögen*]. It is "something intermediate between the imagination properly so-called and the understanding, since the latter is completely active [*tätig*]" (*Tr.*, p. 19–20). Lachterman explicitly equates the faculty of fictions with the faculty of imagination (Lachterman 1992, p. 504). However, since the faculty of fictions does not produce objects given in actuality and since the faculty of imagination produces objects given in actuality, we can safely assert that although these two faculties may share some similarities, they are not the same.

89 "In the conclusion, for example, *a man is an animal*; *Caius is a man*; therefore, *Caius is an animal*, the *major premise* is *intuitive*, as is the *minor premise*, but the *conclusion* is merely thought *symbolically*" (*KdA*, p. 201).

90 "Schließen heißt das *Mannigfaltige der Erkenntniß*, in einer *Einheit des Bewußtseyns* nicht *intuitiv*, sondern durch eine vorhergegangene intuitive Erkenntniß *symbolisch* denken" (*KdA*, p. 199).

gles). That is, *Elements* I.4 is the symbolic conclusion of its proof, serving in the proof of *Elements* I.5 as a given proposition. It is not given in intuition. It is passively presented to our cognition, not actively thought by our thinking faculty. It is grounded in intuition but is not intuition in itself – it is *known*.

The tradition of invention theories is closely related to Euclid's book *Data* and to his notion of given. The propositions in *Data* are based on proving that if something is given, so is another thing given (Taisbak 2003, p. 13).[91] Euclid's *Data* and Maimon's theory and methods of invention are based on the same principle that if something is given then something else is given (or: something is given if we can provide equals to it).[92] Most of the definitions that appear in *Data* are of this form.[93] While the common commentary on the *given* in *Data* is concentrated on givenness in actuality, there is also a more rhetorical interpretation of the given. Such is Gardies' insight, that what characterizes the theory of *given* the most is its grammatical form: if something is given (or some things are given) then another thing is given (Gardies 2001, p. 80). This approach to the given in *Data* corresponds to the notion of given presented above, to include more than being given in intuition alone.

---

91 This notion of being given as the essence of *Data* is also captured in its Hebrew translation as ספר המתנות (freely translated, "the book of gifts"). *Data* was already translated into Hebrew in 1272 by Jacob ben Makhir (Lévy 1997, p. 433) from Hunayn ibn Ishaq's Arabic translation. It continued to be copied in the following centuries (e.g., the Mantova edition, copied in 1612. A copy of the text is found in The National Library of Israel). It is less plausible that Maimon had access to this text in Hebrew. His main source for the text of *Data* was most likely Schwab's German translation from 1780.

92 Maimon does not relate to the notion of *given* in the specific Greek terms *dothen* and *dedomenon*, but only uses the German *gegeben*. Nevertheless, it should be mentioned that *dothen* is used by Pappus to describe something that is admitted, such that if it is possible and producible, so will the sought be possible (Knorr 1993, p. 357). *Dedomenon* is a term that expresses the result of a valid syllogism that was developed by a number of *dothen* (Fournarakis & Christianidis 2006, p. 39). *Dedomena* is also the Greek translation of the title of Euclid's *Data* (see Taisbak 2003).

93 The definitions in *Data* are always in relation to something given, unlike the definitions in *Elements*, which stand by themselves or are in a negative relation (like the definition of the point; see Heath, 1956, Vol. I, p. 158). For instance, *Data*, Def. 1*: "A figure or a line or an angle is given in magnitude if and only if it is possible to provide its equal" (Taisbak 2003, p. 29); Def. 5: "A circle is said to be given in magnitude if its radius is given in magnitude" (Taisbak 2003, p. 34); Def. 6: "And a circle is said to be given in position and in magnitude if its centre is given in position and its radius in magnitude" (Taisbak 2003, p. 34).

Def. 4 is the only one which does not depend on another given: "Given in position is said of points and lines and angles which always hold the same place" (Taisbak 2003, p. 33).

Several kinds of "being given" appear in *Data*, most notably being given in magnitude, in position and in form (Acerbi 2011, p. 121; Taisbak 1991, p. 139).[94] This division could have been helpful to Maimon, were he to use it when working with the concept of *given* in regards to an object given in intuition. Such an example is found in his description of the first kind of analysis. According to Maimon, in *Elements* I.1 ("On a given line AB to construct an equilateral triangle"; *Ueber den Gebrauch*, p. 24), the given line AB is a pseudo-condition and not a true condition. Since by stating the condition "on a given line," we mean an arbitrary assumed line, a determined line whose determination is arbitrary and not necessary (*Ueber den Gebrauch*, p. 24). However, it seems that what is a pseudo-condition is the determined position and determined magnitude of the line, which can be determined arbitrarily. Yet being given the straight line, given in position and in magnitude, is a true condition. Therefore, Maimon's use of *given* in intuition could have benefited were he to adopt the division of *given* appearing in Euclid's *Data*.

---

[94] Ratios are also said to be given, in Def. 2: "A ratio is said to be given for which we can provide the same" (Taisbak 2003, p. 17). In other translations of *Data*, this definition mentions explicitly that ratios are relations between magnitudes, e.g., "A ratio is said to be given, when a ratio of a given magnitude to a given magnitude which is the same ratio with it can be found" (Simson, 1811, p. 367). Acerbi states that ratios are defined separately since they are said to be "the same" and not "equal", i.e. the relation is of identity and not of equality (Acerbi 2011, p. 123). This is in contradistinction to lines, figures and angles who are all said to be given in magnitude if "equals" are provided (see Def. 1).

Chapter 3: Invention, Analysis and Synthesis

Maimon's theory of invention follows the *ars inveniendi* tradition in presenting both general rules of invention and specific methods of invention. When writing on invention in general, Maimon refers to the process of invention as forming syllogism, by which we should seek propositions to serve as premises or conclusions. When referring to invention in a more detailed manner, Maimon presents several notions of invention: invention as logical analysis, analysis in a broader sense and synthesis. Of these three notions, the most important is analysis in the broader sense. It includes different kinds of analysis: ampliative analysis (analysis of the object) and techniques of analysis (e.g. Maimon's methods – the seven kinds of analysis) which cannot be summarized under the title *ampliative analysis*. My aim in this chapter is to create a reconstruction of Maimon's work on invention, analysis and synthesis, while attempting to shed light on some implicit parts of Maimon's work on their relations.

3.1 A General Definition of Invention as Based on Syllogisms: Analysis and Synthesis

The definition of invention recurring the most in Maimon's work on this subject is that of invention as forming syllogisms. Both the genius and the methodical inventor begin with either premises or the conclusion and, by using middle terms, arrive at the sought truth (*Das Genie*, p. 366–367; *Ueber den Gebrauch*, p. 12–13). Although Maimon does not mention it explicitly, this definition of invention as forming syllogism refers to the dual direction of this process: analysis and synthesis.[95] What we assume as implicit in his texts on invention, we find explicit in his book on logic. There, Maimon describes the synthetic method as starting with the premises and using them to arrive at a new conclusion. The analytic method operates in the reverse direction: we begin with the conclusion and from it deduce the already known premises (*Logik*, p. 249–250). This description of the synthetic and analytic processes as reverse is well-known.

---

[95] The *ars inveniendi* tradition gave rise to a fruitful debate over the question of whether invention or discovery is analytic or synthetic. According to the scholastic tradition, for instance, it is synthesis that allows discovery of new things (Timmermans 1993, p. 433) whereas for Descartes, the art of discovery is analytic (Raftopoulos 2003, p. 266). Leibniz chose the golden mean by maintaining that both synthetic and analytic methods can serve invention (Sinaceur 1989, p. 205).

Many philosophers and mathematicians refer to a passage written by Pappus, where he claims that analysis and synthesis are distinct yet closely related processes.[96] It is safe to assume that Maimon was familiar with Pappus' passage (or at least familiar with its spirit), not only because of its popularity but also since it is mentioned in Leibniz's *Nouveux Essais*, a text known to Maimon.[97] Moreover, in *A Short Exposition of Mathematical Inventions,* Maimon mentions Pappus' *Collection* as one of the most essential books in the history of mathematics (*Baco*, p. 266). In the same text, he mentions Plato as the inventor of geometrical analysis [*geometrische Analysis*], referring to it as the art of discovering [*zu finden*] geometric truths by assuming a sought truth as if already true (Maimon uses the verb "invented"; *erfunden*) and by a chain of conclusions arriving at a truth that was already proven using the synthetic method (*Baco*, p. 255).[98] Not-

---

[96] "Now analysis is a way from what is sought, as if admitted, through the things that follow in order, up to something admitted in the synthesis [...] For in the analysis, having hypothesized what is sought as if already in effect, we examine that from which this results and in turn the antecedent of that, until proceeding in this backward manner we come down opposite some one of the things already known or having the order of first-principle. And such a method we call 'analysis', as being like a backward resolution. [...] But in the synthesis, in reverse order, having posited as already in effect what had been obtained last in the analysis, and having ordered in the natural manner as consequents what there [were ordered as] antecedents, and having composed them to each other, we at last arrive at the construction of what is sought. And this we call 'synthesis'." (Pappus, *Collection*, VII, 634; in: Knorr 1993, p. 354).

For the impact of Pappus' text on philosophy see: Hintikka 2012, p. 50; Hintikka & Remes 1974, p. 8–10; Gulley 1958; Raftopoulos 2003, p. 269–270.

However, as Panza's review of the etymology of *analysis* and *synthesis* shows, these are not directly opposite terms. They cannot be narrowed down to passage from the particular to the universal or vice versa. They are better understood as "particular sorts of separation and composition", and include other meanings as well. Among the meanings of *analysis* are "back from solution", "resolution", "towards the solution", "close to the conclusion" and "what brings to unknot something". Among the meanings of *synthesis* are "the act of putting (something) together" and "the act of stating (something) with an accord" (Panza 1997, p. 367).

[97] "L'Analyse des anciens étoit suivant Pappus de prendre ce qu'on demande, & d'en tirer des consequences, jusqu'à ce qu'on vienne à quelque chose de donné ou de connu. J'ai remarqué que pour cet effet il faut que les propositions soyent reciproques, afin que la demonstration synthetique puisse repasser à rebours par les traces de l'Analyse, mais c'est toujours tirer des consequences" (*Nouveaux Essais*, Livre IV., Ch. XVII, § 5: 452).

[98] Another known text referring to the reversed directions of analysis and synthesis is Proclus' commentary of Euclid's *Elements*. Given that Maimon mentions Proclus' work on Euclid on several occasions, it is safe to assume that he was familiar with it. Proclus writes: "[...] For it is necessary to understand generally that all mathematical arguments either proceed from the first principles or lead back to them, as Porphyry somewhere says. And those which proceed from the first principles are again of two kinds, for they start either from common notions and the clearness of the self-evident alone, or from results previously proved; while those which lead

withstanding it seems that when writing on invention, Maimon's inclination is to consider construction of syllogisms as analysis. A short discussion on logical analysis as syllogisms will be presented in Chapter 4 (Section 4.1.7.2). Moreover, before presenting his seven kinds of analysis, he explicitly writes that the more important aspect of invention is analysis: "The entire *art of invention*, as will be shown later on, is based on *analysis*" (*Ueber den Gebrauch*, p. 21; *Das Genie*, p. 375). The following pages will prove this statement to be highly accurate.

3.2 Two Meanings of Invention as Logical Analysis

In *Methods of Invention*, Maimon presents three meanings of *invention:* he refers to the first two notions as logical analysis and the third as synthesis (*Erfindungsmethoden*, p. 139–142). These two kinds of logical analysis may be differentiated as logical analysis and analysis in the broader sense.

In his texts on invention, Maimon repeatedly claims that by *analysis* he means more than just logical analysis. Although he never defines what entails his use of the broad term of analysis as such, exploring his seven kinds of analysis shows us how extensive it is. For instance, he uses the term *analysis* when referring to the invention of new solutions (*Ueber den Gebrauch*, p. 31). Logic, according to Maimon, is unable to provide us with a theory of invention. It is "something so general and undetermined, that we may not hope to take any steps forward with it" (*Das Genie*, p. 371; *Ueber den Gebrauch*, p. 16). He remarks that logic is missing a very important part: it is missing a theory of invention. With a theory of invention it could be used not only as canon, but as organon (*Erfindungsmethoden*, p. 139).[99] Logic may not be sufficient for a theory of inven-

---

back to the principles are either by way of assuming the principles or by way of destroying them. Those which assume the principles are called analyses, and the opposite of these are syntheses – for it is possible to start from the said principles and to proceed in the regular order to the desired conclusion, and this process is synthesis – while the arguments which would destroy the principles are called *reductiones ad absurdum*. For it is the function of this method to upset something admitted as clear" (Proclus, *Commentary on Euclid*, I., ed. Friedlein: 255, 8 – 26; in: Heath 1956, Vol. I, p. 136 – 137).

[99] For Kant, not only can general logic not serve as an organon, it has no role in expanding our knowledge: "Now it may be noted as a sure and useful warning, that general logic, if viewed as an organon, is always a logic of illusion, that is, dialectical. For logic teaches us nothing whatsoever regarding the content of knowledge, but lays down only the formal conditions of agreement with the understanding; and since these conditions can tell us nothing at all as to the objects concerned, any attempt to use this logic as an instrument (organon) that professes to extend and enlarge our knowledge can end in nothing but mere talk – in which, with a certain

tion, but it does play a very important role: Maimon's six kinds of analysis are brought forth to accompany and facilitate the use of logical analysis, not in order to replace it.[100] Maimon asks: "How am I supposed to seek out these [the premises] out of the mass of my knowledge, where the thought premises might be contained?" (*Das Genie*, p. 377; *Ueber den Gebrauch*, p. 22–23), and his answer largely consists of the other six kinds of analysis he presents. It seems that logical analysis is an indispensable component in the process of invention, yet not a sufficient one.

The first kind of logical analysis Maimon mentions in *Methods of Invention* is grounded in logical principles, such as the principles of contradiction and identity:

> Firstly, it [invention] can mean: a *theory of logical invention*, or *logical analysis*, according to which we produce anew various judgments and conclusions by combination [*Combination*]. These, however, are produced only according to *logical principles* [*Grundsätzen*] and are determined for *logical use*. But our *logic* does not lack it. *Lambert's Organon* is very rich in this. (*Erfindungsmethoden*, p. 140)

We may assume that Maimon was referring to logical principles such as syllogism *Barbara* ("all *M* are *C*, all *B* are *M*, all *B* are *C*"), which appears in Lambert's *Neues Organon* (Lambert 1764, § 219, p. 132).[101] For Maimon, what such logical analysis offers to invention is insufficient, since logic supplies only the form (the principles according to which one arrives at new knowledge), whereas the matter is given from another science (*Erfindungsmethoden*, p. 140).[102] Accordingly, in *Methods of Invention* he claims that mathematical analysis turns to geometry for its matter (p. 140). This kind of logical analysis corresponds to Maimon's

---

plausibility, we maintain, or, if such be our choice, attack, any and every possible assertion." (*CpR*, § B86, p. 99). Kant denies that logic takes any part in invention: "Logic, then, is not a general Art of Discovery, nor an Organon of Truth; it is not an Algebra, by help of which hidden truths may be discovered" (Kant 1963, p. 180). Maimon, on the other hand, assigns general logic a role in invention even if, as a form, it depends on other sciences to apply the matter to the process.

100 "However, in my opinion there are *seven* kinds of analysis; so that without the other six, logical analysis could have no use in invention." (*Das Genie*, p. 378; *Ueber den Gebrauch*, p. 24)

101 In *Nouveaux Essais*, Leibniz uses syllogism *Barbara* in order to show how the principle of contradiction is the only principle needed in order to demonstrate the second and third propositions of a syllogism by the first one. He emphasizes the necessity of the use of this principle by geometers in their demonstrations (Livre IV., Ch. II, § 1, p. 328).

102 A similar claim appears in *Logic:* logic is established on general analytic principles (of contradiction and of identity) alone, whereas all other sciences are established on these principles and on specific synthetic principles as well (*Logik*, p. 234–235).

notion of invention as forming syllogisms. In *Logic*, Maimon explains the difference between identity propositions (which do not expand our knowledge) and syllogisms based on categorical propositions (playing an important role in the knowledge expansion): in the case of the proposition "*a* is *a*", the only thing that follows is the same proposition. In the case of "*a* is *a*" and "*a* is *b*", the conclusion "*a* is *b*", is a conclusion which is identical to the second premise. However, when the proposition "*a* is *b*" is connected to the proposition "*b* is *c*", the necessary conclusion is "a is c" and thus is our knowledge expanded (*Logik*, p. 55).

The second kind of logical analysis mentioned in *Theories of Invention* has a confusing title. According to Maimon, we presuppose that some knowledge is given and by analysis arrive at new knowledge. Meaning, this logical analysis is based on hypothetical propositions:

> Secondly, it [invention] can mean: a *logical analysis*, in which we cannot arrive at new knowledge *in itself*, independent of any given knowledge, but can arrive at new knowledge under assumption of given knowledge. Therefore, in this *logical analysis* would be perfectly similar to *mathematical analysis*. (*Erfindungsmethoden*, p. 140)

The second meaning of logical analysis relates to determined objects, such as real objects of mathematics. It is logical analysis that belongs to transcendental logic, not general logic. Maimon equates this kind of analysis with mathematical analysis since both are based on construction. Mathematics assumes *a priori* synthesis and therefore the principle of contradiction does not suffice to determine the objective reality of concepts: there are cases in which we can connect two terms according to the principle of contradiction, but it is impossible to construct both under the same construction. Maimon mentions the *virtuous line* as an example of a connection that is not contradictory (and therefore is possible in a negative way) but that is a *nihil privativum* – an empty object of a concept.[103]

---

[103] In *Essay on Transcendental Philosophy* Maimon defines a possible thing in a negative way as that whose concept is not contradictory. Therefore, it is the most basic kind of possibility and can be related to any possible concept, since whatever withstands the principle of contradiction is possible. A thing is possible in a positive way if it withstands the principle of contradiction and one of the following three requirements: (1) there is an intuition that presents the same relations of the concept; (2) there is an objective ground of possibility, that is, we can think of the subject without the predicate (but not vice versa), so that the synthesis is not arbitrary but is based on the object itself; (3) there is a real definition (*definitio realis*) or an explanation of the way the thing arises (*Tr.*, p. 100–101).

Maimon follows Kant's definitions of *nihil privativum* as an "empty object of a concept" and *nihil negativum* as an "empty object without concept" (defined together with *ens rationis* – an

Equilateral right-angled triangle, on the other hand, is a concept that even though it contains no contradiction, it is impossible to construct the object: it is impossible to construct both determinants, *equilateral* and *right-angled*, under the same construction (*Erfindungsmethoden*, p. 140–142). Maimon mentions that apart from axioms and postulates which are given to our cognition in an immediate manner, mathematical propositions are to be proven, i.e., derived in a mediated way (*Erfindungsmethoden*, p. 142–143). The use of demonstration to arrive at most mathematical propositions might explain why Maimon uses the terms *logical* and *analysis* when referring to mathematics, which is usually regarded by him as synthetic *a priori* science. Attributing logical analysis to mathematics is not trivial. We can assume that by attributing the adjective *logical* to mathematics *(logical analysis* and not merely *mathematical analysis)*, he wishes to emphasize that mathematics is not only productive but also certain [*sicher*], two attributes that, according to Maimon, are required when prescribing a method of invention.

It seems that the importance of logical analysis in the second sense is not damaged by the use of synthetic construction in the process. Even when one needs to conduct a geometrical construction in intuition, it is clear that the construction has the secondary role, whereas logical analysis is the main player. It is clear in several cases. For instance, Maimon writes: "One draws a figure, draws randomly lines in different directions, and waits until luckily those known truths occur in one's memory that apply to the figure and may serve as middle terms to a conclusion" (*Das Genie*, p. 376). First the construction is conducted and without it the logical analysis cannot continue, but the process of invention of a new truth itself is the analysis and not the construction. Also, while Maimon's first kind of logical analysis is equivalent to Leibniz's analysis of notions, his second kind (based also on intuition) is far from it.[104]

Both kinds of logical analysis presented by Maimon are based on syllogistic inference. The first kind of analysis is based on categorical judgments and the second on hypothetical ones. The following segment offers a short examination

---

"empty concept without object" and *ens imaginarium* – an "empty intuition without object") (*CpR*, § B348, p. 295).

[104] Leibniz's analysis of notions is based on identities. It includes simple particles such as "in" and simple complete derived terms such as "*A* in *B*" ["*A dans B*"]. When *A* and *B* are identical, we can substitute one for the other in all the propositions without altering their truth-value (*salva veritate*). Leibniz's example is *triangle* and *trilateral* (Leibniz 1998, p. 207; p. 215). Also, in Leibniz's system *A* is the subject and *B* is the predicate if *B* can be substituted by *A salva veritate* (Leibniz 1998, p. 217).

of the connection between these kinds of judgments and Maimon's notions of invention as two kinds of logical analysis.

3.3 Categorical and Hypothetical Judgments

The first kind of logical analysis is grounded only on the principle of contradiction, whereas the second kind is grounded on construction as well. Although not explicitly mentioned by Maimon, it is implied that categorical judgments belong to the first kind of logical analysis and hypothetical judgments of the type which refers to determined objects (in the form "if a is b then c is d") to the second kind. I restrict my assertion to hypothetical judgments about determined objects only, since for Maimon, there are two kinds of hypothetical judgments: the first is in the form "if a is b, and if b is c, then a is necessarily c". This form is a logical principle, applied only on the object as such, i.e. undetermined logical object. It is based on forms of reason alone (e.g. the principle of contradiction). The second kind of hypothetical judgments is in the form "if a is b then c is d." This kind of judgments refers to determined objects and is grounded not only on the forms of reason but on construction as well (*Logik:* 87–88). Since the first kind of logical analysis refers to determined objects as well, I assign to it categorical judgments (the form: "a is b, b is c, therefore a is c").

Binding categorical judgments with Maimon's notion of logical analysis in the first sense and hypothetical judgments with logical analysis in the second, raises the question: can hypothetical judgments be reduced to categorical judgments, or vice versa? If so, this would suggest that one kind of logical analysis presented by Maimon is more fundamental than the other. The question of transforming categorical and hypothetical judgments is a known one. Bergman's answer is based on Maimon's claim that both hypothetical and categorical judgments share the same form since both express the same one-sided dependence of the predicate on the subject. Therefore, argues Bergman, we can transform every hypothetical judgment into a categorical judgment and vice versa (Bergman 1967, p. 118). He also mentions Maimon's example for transforming a categorical judgment into a hypothetical one (as appears in *Logik*, p. 410): "Instead of saying that "the sum of the angles of a triangle is equal to two right-angles," we can say that "if a certain figure is a triangle, then the sum of its angles is equal to two right-angles" (*Logik*, p. 410). Maimon's example is used by Leibniz in *Nouveaux Essais* to show the opposite, namely, that we can transform hypothetical propositions into categorical ones. Hypothetical propositions, such as "if a figure has three sides, its angles are equal to two right angles", can be transformed into categorical ones by slightly changing the terms, such as: "the angles of every figure with three sides, are equal to two right angles" (*Nouveaux Essais*, Book 4, Ch. XI, § 13, p. 414).[105] However, if we follow Bergman's suggestion, that we can transform every hypothetical into a categorical judgment and vice versa without determining any hierarchy between the two forms, we encounter a problem. Maimon explicitly states in *Logic* that there are propositions that may be transformed from a categorical to a hypothetical form and vice versa, but in essence they are categorical. One example he presents is the proposition "The sum of the angles of a triangle is equal two right angles." Maimon claims that even though this may be transformed into hypothetical form, it is categorical and not hypothetical (*Logik*, p. 395).[106] Another example is the Pythagorean theorem: even though it has the form of a hypothetical judgment, Maimon states that, in essence, it is a categorical judgment (*Logik*, p. 165).[107] Most importantly, when he presents the division of judgments, he mentions that in the narrow sense, categorical judgments include hypothetical ones. Similarly to Kant, Maimon's division includes four forms of judgment: quality, relation, modality and quantity. However, unlike Kant, he maintains that in the narrow sense there are only eight kinds of judgments, two of each form, so that relation judgments in the narrow sense include only categorical and disjunctive judgments. Only in the broader sense his division includes twelve kinds of judgments, adding the hypotheti-

---

105 Leibniz differentiates between conditional and hypothetical propositions (in the form of "if something is given so is something else is given"): In conditional propositions, the subject of the antecedent is the same subject as the subject of the consequent, whereas in hypothetical propositions the subject of the antecedent is different than the subject of the consequent. Eternal truths, writes Leibniz, are all conditional propositions such as "every figure that has three sides, has three angles", and they can be transformed into categorical propositions. Although, Leibniz explains, it is the categorical propositions that are in fact conditional, and not vice versa: "[...] c'est un quoi les propositions categoriques, qui peuvent être enoncées sans condition, quoique elles soyent conditionelles dans le fonds" (*Nouveaux Essais*, Book 4, Ch. XI, § 13, p. 414). Therefore, hypothetical and conditional propositions can be transformed into categorical ones, which in turn are, *de facto*, conditional.
106 "Wie aber die *reine Mathematik hypothetisch* seyn soll, ist mir unbegreiflich; deswegen weil ich anstatt: *die Summe der Winkel eines Dreiecks ist den zweien Rechten gleich*, den Satz so ausdrücken kann: *Wenn eine Figur ein Dreieck ist, so ist die Summe* u.s.w. wird wahrhaftig der Satz nicht *hypothetisch*" (*Logik*, p. 395).
107 "Auch haben wir in der That keine, von den kathegorischen wesentlich verschiedenen, hypothetischen Urtheile. Dieses Urtheil z. B.: wenn ein Dreieck rechtwinklicht ist, so ist das Quadrat der dem rechten Winkel gegenüberliegenden Seite der Summe der Quadrateder übrigen Seiten gleich, hat blos die äußere Form eines hypothetischen Urtheils, seinem Wesen nach aber ist es kathegorisch, und kann auch so ausgedrückt werden: Das Quadrat der dem rechten Winkel gegenüber liegenden Seite, in einem rechtwinklichten Dreieck, ist u.s.w." (*Logik*, p. 165).

cal judgment under the form of relation (*Logik*, p. 54–55).[108] Maimon articulates the difference between categorical and hypothetical judgments using the principle of determinability for both kinds of judgments: categorical judgments express the relation of determinability between subject and predicate where the subject (or determinable) can be thought without thinking the predicate (or determinant), yet the latter cannot be thought without the former. In hypothetical judgments, the antecedent proposition is independent from the consequence whereas the latter is dependent on the former. The difference between these two kinds of judgments is only formal, according to whether the judgments express relations between concepts, objects or propositions.[109] In this regard, there is no difference between categorical and hypothetical judgments (*Logik*, p. 57).[110]

The possible aim for transforming hypothetical propositions into categorical ones is to get one step closer to identities: when analyzing propositions in a Leibnizian system of analysis of notions, our aim is to arrive at identities. Therefore, while Maimon's theory of invention is intended to expand our knowledge and thus places more emphasis on the second kind of logical analysis and hypothetical form of judgments, the more fundamental kind of analysis and form of judgments in Maimon's philosophy are based on the principle of contradiction alone.

3.4 Two Kinds of Analysis in Mathematics

The two notions of invention in *Methods of Invention* are a late conceptualization of ideas that appeared earlier in Maimon's work on invention, mainly in his articles from 1795. In these articles, he describes how logical analysis in its narrow sense is not the only possible kind of analysis that may be used for invention. There, Maimon discusses invention in regards to mathematics and expresses a similar need for a broader definition of analysis, even if he does not present such a definition on that occasion. In order to invent in mathematics, Maimon

---

[108] In his table of judgments, Kant presents twelve kinds of judgments divided into four forms. The third form is relation, expressing three kinds of relations in judgments: between subject and predicate (categorical judgments), between the ground and its consequence (hypothetical judgments) and last, "of the divided knowledge and of the members of the division, taken together, to each other" (disjunctive judgments; *CpR*, § A70/B95, p. 107; § A73/B98, p. 108).

[109] Kant articulates this difference as the difference between relation and proportion: we consider relation between two concepts in the first and relation between two propositions in the second (*CpR*, § A73/B98, p. 109).

[110] "Die *hypothetischen* machen also, in dieser Rücksicht, keine von den *kathegorischen* verschiedene Klasse von Urtheilen aus." (*Logik*, p. 57)

believes there is a need for a concept of analysis broader than the use of syllogisms:

> The entire *art of invention*, as will be shown later on, is based on *analysis*. However, the concept of analysis given by some mathematicians is *too narrow* and does not include everything it ought to include in order to make analysis the source of all inventions. (*Das Genie*, p. 375)

The need for a broader notion of analysis appears in Maimon's criticism of Schwab's definition of analysis, included in his articles on invention from 1795. Maimon quotes from *Thoughts on Analysis*, Schwab's introduction to his translation of Euclid's *Data*, on how we can arrive from proposition A to proposition E and vice versa, by using propositions B, C, and D as middle terms (*Das Genie*, p. 376; *Ueber den Gebrauch*, p. 21–22).[111] He presents Schwab's definition of analysis: "This study of mutual connection or dependency of two propositions or two concepts by means of middle terms or middle concepts is called analysis" (*Das Genie*, p. 376; *Ueber den Gebrauch*, p. 22). Maimon argues that the concept of analysis should include not only logical analysis, but also the kinds of analysis he himself proposes.[112] Furthermore, he remarks that Schwab defines the analytical method but not analysis (*Das Genie*, p. 377–378; *Ueber den Gebrauch*, p. 23).

---

[111] "§ 1. In order to arrive from proposition A to proposition E, I must often think through propositions B, C, D. It can namely happen that in my ideas-system A will not lead immediately to E but to B, B to C, C to D, and D to E, and that I can only arrive from A to E by using these intermediate propositions. § 2. As I may begin with A, and by means of B, C, D, can arrive at E; so I can also begin with E, and search from what it immediately follows, and once I find the sentence D, [I then] search further from what it immediately follows and in this way find proposition C; and I can continue this research, until I arrive at A. In either case I will have found the connection (or the contradiction) of the propositions A and E. This study of mutual connection or dependency of two propositions or two concepts by means of middle terms or middle concepts is called analysis. § 3. Insofar as A is simpler than E, the definition rather applies to the second procedure. Since there E is resolved into A. However, particularly in geometry, we can consider proposition E as already enveloped in proposition A, thus also the first procedure may be called analysis" (*Das Genie*, p. 376; *Ueber den Gebrauch*, p. 21–22).
[112] "First, I remark here: that this definition of analysis by Mr. Schwab concerns only logical analysis, which as *Conditio sine qua non*, in all investigations (since every demonstrative proposition, as a conclusion, must be resolved into its premises); however, this is far from exhausting the concept of analysis as such. And what can it also contribute to *invention*, to give someone the rule: if you want to invent a proposition (its proof), so you must consider the proposition as a conclusion and seek out its premises. However, he will respond: this is precisely the difficulty! How am I supposed to seek out these [the premises] out of the mass of my knowledge, where the thought premises might be contained? – Thus, as long as we don't know more about analysis, we cannot take a single step forwards" (*Das Genie*, p. 377; *Ueber den Gebrauch*, p. 22–23).

Analysis also includes methods that cannot be exhausted into the analytic method, such as analysis of conditions or analysis of the object (*Das Genie*, p. 378). Asking to broaden the notion of analysis beyond the narrow definition of inclusion of a proposition or a concept (for instance, proposition *A*) in another proposition or concept (for instance, proposition *E*), shows that Maimon is not satisfied with the Leibnizian definition of analysis as the "art of finding mediate ideas" (*Nouveax Essais*, Book 4, Ch. 2, § 7, p. 332).

Schwab's choice of the verb *einwickeln* [*envelop*], expressing inclusion, accords with his definition of analysis as inclusion of one concept (or proposition) in another – similar to Leibniz' analysis of notions. For instance, he uses it in his description of proposition *E* as "already enveloped [*eingewickelt*] in proposition *A*" (Schwab 1780, § 3).[113] For Maimon, on the other hand, analysis is not only the envelopment of concepts or propositions from other concepts and propositions, hence he writes that "the *conclusion E* is not *composed* out of the *premises* (as a concept from attributes), but it is merely a *consequence* of them. Therefore *A* is not *simpler* than *E*; they are both equally simple" (*Ueber den Gebrauch*, p. 23).[114] Maimon does not follow up with a presentation of a definition of his broader notion of analysis, but only refers the reader to his seven kinds of analysis which are methods of invention. Reviewing these methods leads to the discovery of a strong resemblance between Maimon's notion of analysis and a method of analysis used by Greek mathematicians called *diorism*. In diorism, a mathematician specifies the limits of a solution, divides the possible cases of the solution and searches for all possible solutions (Saito & Sidoli 2010, p. 582). According to Saito and Sidoli, diorism is a mathematical investigation serving "an important role in the ancient problem-solving art known as the *field of analysis*" (Saito & Sidoli 2010, p. 579–580). I adopt their differentiation between diorism in a broad and narrow sense (the latter named by them as diorism "in its most basic form", p. 580): In the narrow sense, diorism is a part of the proposition defined in Proclus's commentary on the *Elements* I presenting what is sought in the particular instance (Saito & Sidoli 2010, p. 580; Proclus 1789, p. 20). At times, it is also described as a "definition of goal" (Netz 1999, p. 10). In the broader sense, diorism "treats the conditions, arrangements, and

---

113 See also *Das Genie*, p. 376; *Ueber den Gebrauch*, p. 22.
114 This discussion appears only In *Ueber den Gebrauch* and not in *Das Genie*. Maimon describes that in the synthetic method, *A* is considered independent from *E*, whereas in the analytical method, *A* is considered dependent on *E*. In the analytical method, only logical dependence and logical truth are considered. In the synthetic method, we consider both logical and metaphysical truth (*Ueber den Gebrauch*, p. 23–24).

totality of solutions to a given geometric problem" (Saito & Sidoli 2010, p. 579).[115] In a more limited form, the broader notion of diorism appears when Pappus uses the verb *diorisai* in reference to *diorismos* as a definition which is "the preliminary distinction of when, how and in how many ways the problem will be possible" (*Collection*, VII; in: Bernard 2003a, p. 119). It is the action of determining "the possible and the impossible, and, if possible, when and how and in how many ways possible" (Pappus, *Collection*, III, 13–15; in: Bernard 2003a, p. 119). Whether Maimon knew what diorism is or not, we find in his methods of invention a few kinds of analysis similar to it, such as "analysis of the conditions of possibilities of the solution" (first kind of analysis), "analysis of the cases of a problem or a proposition" (third kind of analysis), "analysis of the cases of the solution" (fifth kind of analysis) and "analysis of the various ways according to which a problem can be solved or a proposition proven" (sixth kind of analysis; *Ueber den Gebrauch*, p. 28–31; *Das Genie*, p. 378–381). The use of the methods of invention is one form of a broader notion of analysis. It is different than the notion of analysis in a broader sense that Maimon presents in his *Logic*. Analysis of conditions or simplifying a problem, for example, are different in nature and do not accord with the definition of analysis as educing a predicate from the concept of the object or from the object itself (*Logik*, p. 122). Maimon does not account for the different notions of analysis he uses in the broader sense in relation to mathematics and philosophy. Their common ground is in their definitions that state that they are not based on envelopment alone – the envelopment of concepts or propositions from other concepts and propositions.

Reading Maimon's criticism of Schwab's definition of analysis is misleading. It draws an incorrect picture of Schwab's actual work in *Thoughts on Analysis*. Even though it is true that in the second paragraph Schwab defines analysis in a narrow sense (Schwab 1780, § 2), further on in the text, Schwab's use of the term *analysis* is much broader. He suggests that in order to arrive at the relation between two propositions or concepts, or at the required middle terms or middle concepts, we need to turn to the given object and assist in construction. For instance, in the fifth part of his introduction, Schwab writes that in order to draw a tangent at a given point on a circle, an analysis needs to be conducted. He mentions that we need to find a middle term (unclear which one as he does not specify) in order for the tangent to be constructed (Schwab 1780, § 5, § 2).

---

115 Saito & Sidoli show how diorism examines all possible solutions through an example from Apollonius' *Cutting off a Ratio* 6.4. This practice is used for finding conditions for the solvability of the problem and to show additional cases of the problem (Saito & Sidoli 2010, p. 606–607).

Elsewhere, he states that the first step of preparation for analysis is bringing the given and the sought together, to attempt to find their connection. He does not however, refer to a logical connection, a middle term as a concept, but to a geometrical connection we can see with our own eyes. Even though this connection is later described as a proportion signified symbolically, we must first turn to the object and take its size [*Größe*] and position [*Lage*] into consideration, in order to arrive at the required relation between the constructed geometrical magnitudes (Schwab 1780, § 11).[116] A closer look into Schwab's example: "from a given point on a circle to draw a tangent", reveals that his method of solution is to "determine the tangent out of the size and position of the circle and of the given point, and then I must *explore the middle concept*, by which I will be guided to the tangent, that is, I must analyze" (Schwab 1780, § 5). Analyzing here means more than dealing with conceptual middle terms; it means stretching out to intuition in order to proceed with the proof. This view is quite similar to Maimon's own view of using construction in order to arrive at required middle terms.

The concept of analysis that we can extract from Maimon's work on mathematics is not identical with the concept of analysis in the broader sense presented in his *Logic*. As will be shown in the following sections, the common ground for both notions of analysis in the broader sense is that they are based on more than only the principle of contradiction. His notion of analysis of the object is based on finding new properties in an object by using its construction in intuition. However, his broader notion of analysis in mathematics has a much more extensive span. It includes a great array of analyses, from practices such as verifying whether a condition is necessary or not (Maimon's first kind of analysis), to transforming one object into another using construction (forth kind of analysis).

3.5 Two Kinds of Analysis in Philosophy: Analysis of the Concept and Analysis of the Object

Maimon's critique of the notion of analysis as not exhaustive is aimed not only at mathematicians, but also at philosophers. The need for a broader notion of analysis arises already in *Essay on New Logic* [*Versuch einer neuen Logik*] published

---

[116] Proof and construction are completing processes. In the seventh part of his introduction, Schwab writes that every problem has four parts: a proposition (which asks the question), a construction (which answers), a proof (which shows that the answer is correct) and analysis, without which the construction could not be found (Schwab 1780, § 7).

in 1794. There, Maimon points his arrow of criticism at Kant's concept of analysis and presents a broader definition of analysis: analysis as educing new knowledge not only from a concept, but from an object as well.[117] He writes:

> *Analytic knowledge* is *knowledge* that is educed [*entwickelt*] from a *concept* of an object or from the *object* itself; thus it assumes the concept of the object. (*Logik*, p. 122)

In Maimon's opinion, Kant's notion of analytic knowledge, as that which is educed only from the concept of the object, is too narrow (*Logik*, p. 122). Kant defines analytic and synthetic judgments in terms of whether the predicate is contained [*ist enthalten*] in the concept or not: if the predicate B is contained in the concept A, then the judgment is analytic and if the predicate B is not contained in the concept A, then the judgment is synthetic (*CpR*, § A6/B10 – § A7/B11, p. 48). He refers to all judgments that cannot be thought using identity as synthetic (*CpR*, § A7/B11, p. 48), whereas according to Maimon, thinking identity is not considered as thinking (*Logik*, p. 28). For Kant, analytic judgment is only explicative in that it does not add any predicate that was not already included in the concept of the subject, while synthetic judgment is ampliative in that it adds a predicate to the concept of the subject (*CpR*, § A7/B11, p. 48). What Maimon proposes in his broader notion of analysis is an ampliative notion of analysis.[118] This difference in their approach is found already in the choice of verb to describe the relation between subject and predicate: Maimon uses the verb *entwickeln* [*to educe*] to describe Kant's notion of analytical knowledge, even though Kant tended to use the verb *enthalten* [*to contain*]. Using the same verb as Kant would be problematic out of the context of a predicate contained in the concept of the subject. Therefore, Maimon uses the verb *entwickeln* to describe both his and Kant's notions of analysis.

This broad notion of analysis is equivalent to the second meaning of invention Maimon presents in *Methods of Invention*. From his work on analysis of the object in *Logic*, we can extract two ways to educe new predicates. This is in addition to Maimon's method of analysis of the object mentioned in the two articles from 1795. Maimon himself does not give any account to these three kinds of

---

117 I chose to translate *entwickeln* as *to educe* because of its meaning: to bring out something latent and to deduce in relation to an object. This seems to be the most suitable translation of this verb. The alternatives, such as *to develop*, *to produce* or *to generate*, indicate a change in the subject, a meaning not appropriate to this case of analysis.
118 Maimon's ampliative analysis is not abduction, at least not in the general sense described by Hintikka as "an inference to the best explanation", explaining a truth by deriving it from an assumed theory together with other sought contingent truths (Hintikka 1998, p. 506–507).

"analysis of the object". First, we should distinguish between analysis of the object based on educing the property immediately from the object (as in the example "a triangle has three angles") and between analysis of the object based on using demonstration to arrive at the new predicate (as in the example of the Pythagorean theorem). The method of analysis of the object is based on transforming the given object into the sought object.

3.5.1 Analysis of the object: educing a predicate immediately from the object – the case of educing three angles

The first kind of analysis of the object that can be extracted from Maimon's work is that of educing a predicate immediately from the object. This notion appears in his discussion of identities and analytic and synthetic thought in *Logic*. According to Maimon, the thought of identities in which the predicate is educed from the concept of the subject (often considered by philosophers as analytic thought) is in fact no thought at all. Because the predicate is already thought in the concept of the subject, there is no action of thinking a new thought (*Logik*, p. 28 – 29). This accords with Maimon's definition of thinking as an active action of the understanding, which thinks an object by thinking the rule by which it arises (*Tr.*, p. 32 – 33). Maimon calls *analytic thinking* the case where the predicate is educed from the subject itself by thought, whereas in *synthetic thinking* the predicate cannot be brought out by educement [*Entwickelung*] of the subject (*Logik*, p. 29). He offers the following examples to illustrate the difference between analysis (in its broader sense) and synthesis:

> This proposition, for instance, "*a triangle* (space enclosed by three lines) *has three angles*", although *three angles* is not contained in the *concept* of the triangle, is nevertheless *analytic*, because the *three angles* appears out of its educement [*Entwickelung*], what is contained in the triangle itself. But this proposition: "*a triangle can be right-angled*" is *synthetic*, because *being right-angled* cannot arise from the *already accomplished* [*vollbrachten*] construction of the triangle in general, but rather [it arises] from the *new* construction of the *right-angled triangle*. (*Logik*, p. 29)

Given a triangle, we can arrive at the new unknown proposition "a triangle has three angles" by turning directly to the object and not to the concept of the object (*triangle*), and by educing from it the new property *three angles*. Though Maimon specifically names such a process *analysis*, he does not explain how this analysis is in fact conducted and what justifies such a title.

The change in Maimon's approach to naming the arrival at the new property *three angles* as *analysis* or *synthesis* occurred around the time of the publication of his *Logic*. In *Essay on Transcendental Philosophy* (1790), he refers to the proposition "A triangle (a three-sided figure) has three angles" as an example of a proposition which is not a logical truth (as declared by Wolff), but as a synthetic truth to which we can assign only subjective necessity (since we do not have the ground for claiming its objective necessity) (*Tr.*, p. 404–405). Moreover, in *Giv'at Hamore* (1791), Maimon states that *three angles* is a property [תאר העצם; *Eigenschaft*] of the triangle which is necessarily derived from the subject. It is not a determinant [חלק העצם; *Bestimmung*] of the subject, i.e., it is not part of the definition of the triangle. *Property* is inferred from the subject but is not part of its definition. Determinant is part of the essence. Therefore, *three angles* is not a determinant of *triangle* but a property, whereas *three lines* is a determinant of *triangle* and not a property ["keine Eigenschaft sondern eine wesentliche Bestimmung"] (*Giv'at Hamore*, p. 76). In his translation of Aristotle's *Categories* (published after *Logic*, in 1794 as well), Maimon presents the proposition "a triangle has three angles" as an example of a synthetic judgment (*KdA*, p. 194). Similar to his statement in *Logic*, he claims that the predicate is not contained in the concept of the subject, but in the subject itself. The difference is in assigning the proposition once as synthetic and once as analytic, while the two texts were published roughly at the same time. This change in Maimon's assignment of this example to analysis or synthesis does not mark a change in his description of the example, but rather a change in his concept of analysis. This change has been neglected by previous commentators, who have either tried to prove that the property *three angles* can be analytically derived as a conclusion from premises that are grounded in intuition (Freudenthal 2006, p. 89),[119] referred to analysis of the object as synthesis (Atlas 1964, p. 243),[120] or neglected this

---

119 Freudenthal mentions this example as an example of analytic judgments that are based on the subject and not on the concept of the subject. However, he only refers to the kind of analysis of the object that is indirect, i.e. inference of a conclusion derived analytically from premises. He shows how, by using propositions such as Postulates 1 and 2 together with Definition 8 of Book I of *Elements*, we can infer that a triangle has three angles. He then concludes: "This is conceptual, analytic thought that is independent of intuition and yet enlarges our knowledge" (Freudenthal 2006, p. 89–90). However, what I refer to as the first kind of analysis of the object (extracting the new property directly from the object) is an action of judgment and not of inference – a judgment that is in itself grounded in intuition.
120 When Atlas refers to Maimon's analysis of the object and the example of discovering the three angles of the triangle, he writes that the connection between the two concepts is "not analytically derivable" (Atlas 1964, p. 243), thus relating to Maimon's narrow definition of analysis as the only definition of analysis. In describing Maimon's three classes of propositions, Atlas mentions analytic propositions governed by the principle of contradiction, synthetic propositions *a posteriori* (about objects of experience) and synthetic propositions *a priori* (about math-

3.5 Two Kinds of Analysis in Philosophy

broader notion of analysis as analysis of the object all together (Bergman 1967, p. 109).[121]

Much of the confusion involved in referring to the three angles as being educed from the object itself is due to the question of the necessity of the connection between the subject and predicate. On the one hand, as mentioned by Freudenthal, Maimon sets out against assigning objective necessity to a connection that is not clear to the understanding (as it is not based on the principle of contradiction) (Freudenthal 2006, p. 89).[122] On the other hand, when in *Logic*, Maimon discusses the grounds for different kinds of judgments, he assigns necessity to the connection of subject and predicate in analytic judgments of the kind of analysis of the object (together with the example of the three angles). According to Maimon, the ground for identity judgments is in the concept of the object. He mentions the proposition: "a regular decahedron is regular" as an example: since the predicate is thought in the concept of the subject, its ground is in the concept. When the connection between the subject and predicate is in the object and not in the concept of the subject, and this connection is necessary, then the ground is in the object. Maimon suggests the proposition: "a three lateral figure has three angles" as an example. Its ground is not in the concept of the subject since the predicate *three angles* is not contained in the concept of *three lateral figure*, but only in its construction. When the connection between the subject and predicate is only possible, the object is a consequence of this

---

ematical objects), stating that the latter are dependent on the analysis of the object (Atlas 1964, p. 244).

121 Bergman refers to two notions of analysis that Maimon presents but discusses only the definition relating to the principle of determinability and neglects to mention the notion of analysis of the object. Bergman writes that Maimon distinguishes between these two forms of analytical judgments: '(*a*) an analytical judgment that expresses the relation of the determined to the determinable and to the determinant, e.g. "the right angle is an angle," "the right angle is right"; (*b*) an analytical judgment that expresses the relation of the determinant to the determinable, e.g. "white is a color"; analytical judgments of this kind are subject to the principle of determinability and are not possible without it' (Bergman 1967, p. 109). However, Maimon's notion of analysis of the object refers to the relation between the determinant and the determinable, but this knowledge can be derived not only from the concept but from the object as well.

122 "The Understanding prescribes the productive imagination a rule to produce a space enclosed by three lines. The imagination obeys and constructs the triangle, but lo and behold! Three angles, which the understanding did not at all demand, impose themselves. Now the understanding suddenly becomes clever since it learned the connection between three sides and three angles hitherto unknown to it, but the reason of which remains unknown to it. Hence it makes a virtue of necessity, puts on a imperious expression and says: A triangle must have three angles! – as if it were here the legislator whereas in fact it must obey an unknown legislator" (Maimon, *Antwort*, GW III, p. 198–199 in: Freudenthal 2006, p. 87–88).

connection. In this case, Maimon mentions the proposition "a triangle can be right-angled" as an example. Its ground is not in the concept of the object or the object, but this connection itself is the ground of a new object: a right-angled triangle (*Logik*, p. 51–52).[123] We can only assume that Maimon intended to assign subjective, and not objective, necessity to analytic judgments of the kind of analysis of the object. It seems that the broader notion of analysis as analysis of the object is meant to provide an analysis that is both productive and certain. As such, it can be helpful in a theory of invention. The problem that arises with this kind of analysis of the object (based on educement of a new property by turning directly to the object) is that the ground for connecting predicate and subject is left opaque to reason. It does not offer any method to arrive at new knowledge other than "looking at the object" and searching the sought predicate by using the faculty of imagination. In this, Maimon's method is no different than Schwab's suggestion to look for the connection between the given and the sought that "lies in front of our eyes" ("wenn man *beyde vor Augen liegen hat*"; Schwab 1780, § 10).

Maimon does not offer an explanation for naming the process of extracting a new property out of an object as *analysis*. Nevertheless, justifications may be found elsewhere in his work. We can identify two kinds of explanation: the first is based on the type of determination and it explains the immediate educement of a predicate; the second is based on conclusions of hypothetical judgments and it explains the indirect educement of a predicate (therefore its description appears in the next section). When in a determination, *b* is a possible determinant of *a*, a new object arises. Maimon's example is the judgment: "a triangle can be right-angled". Being right-angled is a possible determination of a triangle, and a new object (a right-angled triangle) arises. When *b* is a necessary determinant of *a*, a new object does not arise, but *a* and *b* are connected in an object. Maimon's example is the judgment: "a triangle has three angles". If *b* is contained in the concept *ab* and not only in the object, then this connection between *a* and its determinant *b* is *a priori* objective, since what is considered as necessary for the concept is necessary in regards to the object

---

123 According to Maimon, the concept *right-angled triangle* originates in the judgment: "a triangle can be right-angled" since the consciousness of the subject preceded the consciousness of the predicate. In both the concept and the judgment, the subject and predicate are the same. In analytical judgments which are identities, the consciousness of the subject does not precede the consciousness of the predicate and then it is not thought. Therefore, the judgments "a right-angled triangle is a triangle" and "a right-angled triangle is right-angled" are not thought. But in the judgment "a triangle can be right-angled" the consciousness of the subject precedes that of the predicate, which cannot be thought by itself (*Logik*, p. 26–28).

as well (*Logik*, p. 50–51). As a result, analysis of the object can be defined as "finding that *b* is a necessary determinant of the subject *a*, when their connection is in the object". If their necessary connection is in the concept, then it is analysis of the concept. Both processes are named *analysis* since they dissolve [*auflösen*] a necessary knot. This is contrary to synthesis, where the connection between determinant *b* and determinable *a* is not necessary but only possible. In this case, the process is called *synthesis* since we compose the determinant with the given determinable. However, what is lacking is an explanation of how we arrive at the knowledge that a determination of an object that is not a part of its concept is necessary.

3.5.2 Analysis of the object: educing a predicate indirectly by using demonstration – the case of the Pythagorean Theorem

In the second kind of analysis of the object, the educement of a new property from the object is not immediate but indirect, using demonstration. This difference between the two kinds of analysis of the object does not reside in Maimon's description of analysis of the object but is implicit in the examples he uses. When Maimon describes the various kinds of knowledge (analytic, synthetic, analytic-synthetic), he refers to the Pythagorean Theorem (*Elements* I.47) as analytic knowledge:

> Therefore I expand the explanation of *analytic knowledge* further, as I understand it, as produced [*hervorgebrachte*] knowledge *by educement* [*Entwickelung*] *in general* (of the concept or the object), so that according to my assertion, therefore, not only the attribution of already *thought essential* determinations to the *concept* of the object belongs to *analytic knowledge*, but even the attribution of the *properties* of this essence. On the other hand, I narrow *synthetic knowledge* only to knowledge in which a *real object* is determined. The concepts of the objects of *mathematics* (produced by thought according to the principle of determinability) belong to *synthetic knowledge*. For instance, the concept of a triangle arises [*entsteht*] owing to the synthetic judgment: "*space can be enclosed by three lines*". However, *the propositions* in which *properties* are attributed to an object, belong to *analytic knowledge* although they are not educed [*entwickelt werden*] from the *concept* but rather from the *object*, such as, for instance, the *Pythagorean Proposition*. But this deviation from *Kant* concerns only the *nominal definition* [*Worterklärung*]. However, on the matter itself we are in complete agreement. (*Logik*, p. 123–124)[124]

---

[124] Analytic-synthetic knowledge is defined by Maimon as "knowledge of possible *relations* that refer to an *object as such*" (*Logik*, p. 122). This refers to general forms of thought that are relation-concepts, such as cause and effect, where one cannot be thought without the other (*Logik*, p. 124).

Maimon renames knowledge that in Kantian terms is synthetic – finding new properties of an object not based on the principle of contradiction alone but also based on the construction of the object in intuition – as analytic. To the realm of synthesis, he leaves only the creation of new objects. Thus, a triangle is created as a new object by synthesis and by analysis of the object we find that it has three angles. By adding the determination *right-angled* as a dependent predicate to the subject *triangle,* we create the new real synthesis *right-angled triangle*. It has a new consequence: the Pythagorean theorem.[125] By referring to an action that involves not only thought (which is based on the principle of contradiction alone), but also cognition (based on the principle of contradiction as well as on intuition), Maimon presents a notion of analysis that includes not only analytic thought, but also analytic cognition. Stating that he is in complete agreement with Kant only to present new definitions, Maimon *de facto* revokes this statement. For changing all the definitions of analysis and synthesis infers a new philosophical system, different than Kant's.

Although not mentioned by Maimon, this kind of analysis of the object differs from the first in that the connection between the subject and predicate is indirect: we need to use inference to arrive at the sought knowledge. The relation of determinability here is not immediate, but indirect. To better understand the difference between the two suggested kinds of analysis of the object, we can turn to Maimon's explanation of two kinds of conclusions. Conclusions are determined by Maimon as a relation of the given manifold to the unity of consciousness that is not immediate but indirect and not intuitive but symbolic (*Logik*, p. 72–73). Maimon writes that the relation between the subject and predicate in the conclusion is not immediately known from their relation of determinability – we must first turn to the knowledge that arises immediately from the relation of determinability of the subject and another predicate ("a is b"). Subsequently, we turn to consider the predicate as a subject and its immediate relation to another predicate ("b is c") and finally we attribute the first subject with the predicate which is not connected with him in a direct relation of determinability ("a is c") (*Logik*, p. 73). In addition, he mentions that immediate conclusions are judgments where the connection between subject and predicate is immediately recognized, so that they are indemonstrable. Hypothetical princi-

---

[125] For a synthesis to be determined by the principle of determinability as real, it has to withstand three conditions: (1) dependence of the predicate on the subject; (2) presentation of the object in intuition; (3) produce new consequences (*Tr.*, p. 88; p. 391).

According to Yakira, Maimon's notion of 'determination' was influenced by Leibniz to such an extent that some sentences echo Leibniz' text. The notion of the 'determinable', however, is a complete innovation offered by Maimon (Yakira 2003, p. 63–64).

ples are immediate conclusions and hypothetical judgments are indirect conclusions (*Logik*, p. 82). Connecting subjects and predicates immediately is judging and is an action of the faculty of understanding. Connecting subjects and predicates in an indirect manner is inferring and is an action of the faculty of reason (*Logik*, p. 209). Therefore, the two kinds of analysis are conducted by different faculties: the first is an act of the understanding whereas the second is an act of the faculty of reason.

3.5.3 The method of analysis of the object: transformation of the given object – the case of *Elements* I.5

In his two articles on invention published in 1795, Maimon mentions analysis of the object as a central method of invention (*Ueber den Gebrauch*, p. 30; *Das Genie*, p. 380). Maimon does not account for the difference between analysis of the object as presented in his theory of invention and as it appears in his work on logic. The first refers to a specific method which is based on the transformation of the object (a discussion on this method is presented in section 4.1.4). Maimon presents *Elements* I.5 as an example of his method of analysis of the object: to arrive at the sought conclusion, we need to alter the given object into another object so that we can apply given propositions on it and thus arrive at the sought proposition (*Ueber den Gebrauch*, p. 30). Analysis of the object as it appears in *Logic*, however, does not require such a transformation of the object in order to conduct analysis. Indeed, in the specific example of the Pythagorean theorem we alter the given object by using construction. Nevertheless, there are other cases, such as *Elements* I.4 or *Elements* I.15, where arriving at a new property of the given object is attained by using demonstration but without any alteration of the object.[126]

When writing on the method of analysis of the object, Maimon claims that what is important in invention is not the analyzed object but the analysis itself, how it is conducted step by step (*Ueber den Gebrauch*, p. 32). In this aspect, the immediate analysis of the object differs from the method of analysis of the object

---

[126] *Elements* I.4: "If two triangles have the two sides equal to two sides respectively, and have the angles contained by the equal straight lines equal, they will also have the base equal to the base, the triangle will be equal to the triangle, and the remaining angles will be equal to the remaining angles respectively, namely those which the equal sides subtend" (Heath, 1956, Vol. I, p. 247).

Elements I.15: "If two straight lines cut one another, they make the vertical angles equal to one another" (Heath, 1956, Vol. I, p. 277).

and from the analysis of the object conducted using demonstration. Whereas the first kind of analysis of the object (judgments) is the action of the understanding and the second kind (inference) the action of reason, the third (alteration of the object itself) is an action carried out by the faculty of imagination.

3.6 Invention as Synthesis

The status of synthesis in Maimon's work on invention is intriguing. On the one hand, different kinds of synthesis are mentioned by Maimon throughout his work as highly productive means of invention. Such means are the method of fictions (e.g. *Prg.*, p. 54–55). However, in Maimon's works dedicated to invention, the importance of synthesis seems much smaller. Synthesis is not mentioned in the two articles on invention published in 1795 where he explicitly states that invention is based on analysis (*Ueber den Gebrauch*, p. 21; *Das Genie*, p. 375). In *Methods of Invention*, however, synthesis is mentioned as a third meaning of a theory of invention, alongside the two notions of analysis (*Erfindungsmethoden*, p. 143). The meaning of a theory of invention as synthesis is based mainly on the methods of conversion and of assuming a problematic proposition as true. Thus, it is probable that the change that occurred between the publication of the two articles in 1795 and the time when Maimon wrote *Methods of Invention* is due to the improvements he made in these two methods of invention (see Sections 4.2 and 4.4). Another method that is a method of synthesis, but one that he does not explicitly mention as such, is the method of generalization (Section 4.3).

In *Methods of Invention* he describes invention as synthesis in the following way:

> However, a *theory of invention* can also have a third meaning. It can mean a theory by which we can arrive at new synthetic knowledge from *given synthetic knowledge*, according to *principles* of *synthetic knowledge in general*. But I cannot prove that such a theory is *possible* better than by the *act* itself, as I want to put to the test here a few propositions from this theory (which shall be first postulated). (*Erfindungsmethoden*, p. 143)

The most notable aspect of this definition is that invention as synthesis begins with something given. By introducing this definition, Maimon excludes other productive kinds of synthesis that may serve invention but do not begin with anything given, such as the method of fictions. This definition accords with the definition of invention as something that begins with the given and as forming syllogisms. It also accords with Maimon's famous claim that our cognition begins with something given and not with ideas such as the absolute unit and infinite in itself.[127] When writing about invention as synthesis, Maimon does not mention his principle of determinability, a criterion to verify whether a synthesis of a subject and a predicate is real thought (and not formal or arbitrary thought; *Logik*, p. 310). This kind of synthesis, where an absolute concept in which the subject is something that can be thought in itself without the predicate (*Tr.*, p. 84), also begins with something given: the determinable. For instance, in the determination *equilateral triangle*, the determinable *triangle* is given before the determinant *equilateral*. Moreover, real thought is distinguished from formal and arbitrary thought in that it contains both form and a given object (*Logik*, p. 310). In constructing an equilateral triangle as an object given in intuition in *Elements* I.1, we also begin with something *given*, in this case – a straight line.

Maimon's principle of determinability was not included in the scope of his published works on invention. Most probably because it is a principle that verifies whether a given synthesis is real thought, not one that generates a new synthesis. He presents three criteria which a real synthesis should withstand. It needs (i) to be an absolute concept, (ii) be constructed as an object in intuition and (iii) produce new consequences. The final criterion is problematic. The criterion that real synthesis produces new consequences (or "conclusions" [*Folge*]; see Schechter 2003, p. 23) is problematic since it negates reality from some synthesis given in actuality and that bares no new consequences. Consequently, in a sense it equates reality and production of new consequences. Let us consider the following case: *right-angled trapezium* is an absolute concept and can be constructed as an object in intuition. As such, it fulfills two of the three requirements for a synthesis to be real thought. It does not, however, have new consequences. It is not arbitrary thought given that it not only has an object given in intuition, but also is an absolute concept. As well, it is not only formal, since an object is given in intuition. *Isosceles trapezium*, in comparison, is an absolute concept, can be constructed as an object in intuition and has a new consequence: its diagonals are equal. By determining the first synthesis (*right-angled trapezium*) as not real and the second synthesis (*isosceles trapezium*) as real, a difficulty arises: it equates real synthesis with producing new consequences and denies reality from absolute concepts that have an object given in actuality. This difficulty could have been avoided if we neglected the third criterion of pro-

---

[127] "But we should note that both the primitive consciousness of a constituent part of a synthesis (without relating this part to the synthesis) as well as the consciousness of the complete synthesis are mere ideas, i.e., they are the two limit concepts of a synthesis, in that without synthesis no consciousness is possible, but the consciousness of the completed synthesis grasps the infinite itself, and is consequently impossible for a limited cognitive faculty. [...] So we start in the middle with our cognition of things and finish in the middle again" (*Tr.*, p. 349–350).

ducing consequences. Additionally, Maimon's example of an absolute concept is problematic. He defines an absolute concept as a synthesis where one part of it can be thought in itself (the subject), whereas the other part of the synthesis cannot be thought without the subject (the predicate). As an example, he mentions that one cannot think *right-angled* without thinking *triangle*, and therefore *right-angled triangle* is an absolute concept (*Tr.*, p. 84).[128] However, Euclid's tenth definition in *Elements* presents the right angle between two straight lines as a concept and as an object, without the thought or presentation of a triangle.[129] Therefore, *right angle* can be thought and constructed in intuition without the determinable *triangle*.[130] Furthermore, we can think and present in intuition a right-angled trapezium without thinking or presenting in intuition a right-angled triangle. Although every quadrature can be divided into triangles, dividing a right-angled trapezium (a form which has two right angles) given in intuition into triangles can be done using only one right-angled triangle at a time. On the conceptual level, we do not use *triangle* in order to define *trapezium*. Hence, *right-angled trapezium* is not in a relation of determinability or dependence with *right-angled triangle*. To avoid eliminating the use of an absolute concept as a requirement of real synthesis, we can offer one of two options: either eliminate the requirement that each predicate has only one subject (a requirement presented in *Tr.*, p. 380), or claim that being right-angled is a property of *figure*, or even, in the case of a right angle (as an object presented in itself), of *space* itself. Melamed mentions that two of the three proofs provided by Maimon to support his claim that a predicate cannot belong to more than one substance, are not very convincing. The third proof is based on *reductio ad absurdum* that assumes two independent concepts (neither of which is a predicate of the other). If we assume that they share a predicate, then consequently there would be a necessary connection between them, thus contradicting the assumption (Melamed 2004, p. 91– 92). However, this is the case presented above: *right-angled triangle* and *right-angled trapezium* share a predicate, and neither is a predicate of the other. However, they are connected in that both have *space* as a determinable.[131] Then again, having a shared determinable does not contradict

---

128 Maimon mentions the absolute concept as a criterion of real synthesis in *Tr.*, p. 91.
129 *Elements*, *Def.* 10: "When a straight line set up on a straight line makes the adjacent angles equal to one another, each of the equal angles is *right*, and the straight line standing on the other is called *perpendicular* to that on which it stands" (Heath, 1956, Vol. I, p. 153).
130 Maimon refers to this example also in terms of relation of determinability: *triangle* is the determinable and *right-angled* is its determination (*Tr.*, p. 145).
131 For *right-angled triangle*, the order of determination is *space-figure-triangle-right-angled*, while for *right-angled trapezium*, it is *space-figure-quadrature-trapezium-right-angled*. Therefore,

the assumption that one is not a predicate of the other. Therefore, we do not arrive at a contradiction. This shows that Maimon's third proof is lacking as well. A better example of an absolute concept that Maimon could have used is the example "space enclosed by three lines", used in *On Symbolic Cognition*. The determinable *space* can be thought of in itself and yet we cannot think *three lines* without *space* (*Tr.*, p. 284).

Out of the three criteria presented by Maimon to determine whether a synthesis is real thought, we should remove the last and remain with only the first two criteria: an absolute concept constructed as an object in intuition. There is need for both criteria since, as Maimon claims, an absolute concept can be only formal, not real (such as a regular decahedron; *Logik*, p. 312) and since presenting an object in actuality does not indicate it is a real synthesis of the understanding, only that it is a real synthesis of the imagination in time and space (*Tr.*, p. 105). The second criterion may be change into presenting a *new* object in intuition, not just *an object in intuition*. This change of the criterion accords with the definition of synthesis as the introduction of a new object (*Logik*, p. 29). It also withstands cases such as *right-angled trapezium*, that may not have new consequences, but do involve the construction of a new object.

The most interesting notion of invention as synthesis is found not in the works dedicated to invention, but rather in his book on logic. There he defines synthesis as the introduction of a new object (*Logik*, p. 30; p. 123). The definition of synthesis as the introduction of a new object is more general than its definition based on relation of determinability. For instance, the caustic curve (mentioned in Chapter 2 as an example of invented mathematical object) accords with the definition of introducing a new object, yet does not comply with the definition of synthesis based on relation of possible determination.[132] In the following section, synthesis in the narrow sense will be compared to the notion of analysis of the object. What is considered by Kant as synthetic *a priori* judgments, is divided by Maimon into two types of knowledge.

---

trapezium is not a predicate of *triangle*, nor is *triangle* a predicate of *trapezium*, yet both share a predicate and a determinable.
132 A caustic curve is a synthesis of a plane with caustic surface. A plane is a "surface which divides space into two congruent parts" (Heath, 1956, Vol. I, p. 176). A caustic surface envelops reflected or refracted rays of light by a curved surface (when Maimon describes this curve, he uses the term *concave mirror* [*Brennspiegel*]; *Baco*, p. 299). Therefore, *plane* and *caustic surface* are two disjunctive determinants of the same determinable *surface*. That is, *caustic curve* is a new object that cannot be reduced to a mere determinant of either *plane* or *surface*.

3.7 How are Synthetic *a priori* or Ampliative Analytic Judgments Possible?

By introducing new definitions of synthesis and analysis of the object, Maimon divides Kant's synthetic *a priori* knowledge into two types of knowledge. For Kant, both propositions "given three straight lines, a figure is possible" and "the straight line between two points is the shortest" are synthetic *a priori* (*CpR*, § A47/B65, p. 86; § B16, p. 53). According to Maimon's definitions, however, only the first is considered synthesis. Following the presentation of his new definitions, Maimon rephrases Kant's question on the possibility of synthetic *a priori* judgments to also include judgments that are analytic and ampliative:

> Regardless of the difference between the Kantian explanation of the synthetic judgments and mine, we are in agreement on the matter itself, except that according to me, the question: *how are synthetic a priori judgments possible?* must be expressed exhaustively in this way: how are synthetic, or analytic *judgments, whose predicate is not thought as contained in the concept of the subject, but rather in the subject itself, possible?* (*Logik*, p. 258)

By renaming a group of propositions, which are called *synthetic* by Kant, *analytic* (i.e. ampliative analytic judgments), Maimon accords with the definitions of *synthesis in general* as unity in manifold (*Tr.*, p. 20; *CpR*, § A77/B103, p. 111) and of *analysis in general* as manifold in unity.[133] Before addressing the renaming of some synthetic *a priori* propositions as ampliative analytical, it should be clarified that whether a proposition is regarded as ampliative analytic or synthetic, the question of the possibility remains the same. Since, in either case, the propositions are not absolutely *a priori*. This is why, one may assume, Maimon does not use the term *a priori* in his own version of Kant's question. Maimon presents two definitions of *a priori*: absolutely *a priori* is a cognition that precedes the cognition of the object and is grounded on the principle of contradiction alone. *A priori* in the broader sense is a cognition that precedes any sensation.[134]

---

[133] As mentioned by Kant, defining synthesis as combination is based on the concept of unity of the manifold (*CpR*, § B130–131, p. 152). Therefore, unity in manifold is the most general definition of synthesis. Accordingly, analysis is defined as dissolution [*Auflösung*] and it is based on the more general concept of manifold in unity.

[134] "The absolutely *a priori* [*a priori absolut betrachtet*] is, for *Kant*, a type of cognition that must be in the mind prior to any sensation. For me, on the other hand, the absolutely *a priori* is a type of cognition that precedes cognition of the object itself, i.e. [it is] the concept of an object in general along with everything that can be asserted about such an object, or [a type of cognition] in which the object is only determined by means of a relation, as for example the objects of pure mathematics" (*Tr.*, p. 168–189). Maimon's definitions of *a priori* and absolutely *a priori* are distinguished from Kant's also in that they are not based on the notion of necessity.

3.7 How are Synthetic *a priori* or Ampliative Analytic Judgments Possible?

The definition of absolutely *a priori* postulates that when the cognition of the objects is prior to the cognition of the relation between the objects, the cognition is *a posteriori* (formally, not materially).[135] The two kinds of *a priori* were presented by Maimon already in his first book, where he kept the "traditional" differentiation between analytic knowledge as grounded in the principle of contradiction alone and synthetic knowledge as grounded on intuition as well. Consequently, cognition of determined objects given in intuition was considered synthetic. In *Logic*, on the other hand, by introducing the notion of ampliative analysis, he presents a new kind of analysis that is formally *a posteriori* (in the narrow sense) and based on intuition. Therefore, Maimon's new version of the question is about the possibility of *a priori* judgments that are not absolutely *a priori*. Furthermore, contrary to analysis in the narrow sense, ampliative analysis is not pure, since for Maimon, pure is "a product of the understanding alone (and not of sensibility)" (*Tr.*, p. 56). Thus, ampliative analytic judgments are not absolutely *a priori* and not pure. Similar to synthetic *a priori* judgments and contrary to analytic judgments in the narrow sense.

Maimon presented the altered version of Kant's question only in his notes and clarifications, not as a central text of his *Logic*. It was a by-product of his thoughts on real analytic and synthetic thought and he did not elaborate on

---

Kant defines *a priori* in relation to experience (as what is independent of all experience; *CpR*, § B2, p. 42–43) and in relation to necessity: an *a priori* judgment is "a proposition which in being thought is thought as *necessary*" and absolutely *a priori* is a proposition which "is not derived from any proposition except one which also has the validity of a necessary judgment" (*CpR*, § B3, p. 43). According to Senderowicz, Kant's definition of *a priori* as a cognition independent of any experience is derived from his notion of an *a priori* judgment as a necessary judgment (Senderowicz 2003, p. 180).

135 "So, taken in its strictest sense, absolutely *a priori* cognition [*Erkenntnis a priori im engsten Verstande und absolut betrachtet*] is the cognition of a relation between objects that is prior to the cognition of the objects themselves between which this relation is found. Its principle is the principle of contradiction (or identity). But if the cognition of the object must precede representation of the relation, then, in this strict sense, it is termed *a posteriori*. From this it follows that we do not have *a priori* cognition of axioms of mathematics, that is to say, they are not *formaliter a priori*, although they are *materialiter* (in time and space) *a priori*" (*Tr.*, p. 169). A similar claim is found in *Logic*. Maimon maintains that mathematical cognition is not *a priori* cognition in the narrow sense, since it is not a formal cognition that refers to the object in general, but a real cognition that refers to determined objects (*Logik*, p. 117).

I believe that when Maimon speaks of absolutely *a priori* analytic propositions (referring to the object in general), he refers to universal propositions such as "*ab* is *a*" (e.g. *Logik*, p. 413) rather than as suggested by Bransen, to particular propositions where the relation is between certain determinable and determinant (Bransen 1991, p. 69).

74 — Chapter 3: Invention, Analysis and Synthesis

its implications.[136] However, we can divide Kant's synthetic *a priori* judgments into two types of judgments, according to whether the relation of determinability between subject and predicate is necessary or only possible and proven not to be necessary.[137] Considering the different modality of these judgments[138], Kant's synthetic *a priori* judgments can be divided into:

(a) Synthetic judgments, such as "a triangle can be right-angled": the relation of determinant to the determinable is of possibility. A new object arises when a determinant is a possible determinant of a determinable. These judgments are based on absolute concepts where the subject (or determinable) can be thought in itself and the predicate (or determinant) can be thought only in relation to the subject. In other words, there are several possible determinants for the determinable so that the relation between the determinable and determinant is only possible and not necessary. These are problematic judgments ("*a* is *ab*"). They cannot be transformed into judgments which are apodictic, where the predicate is necessarily contained in the concept of the subject ("*ax* is *a*") without changing the subject and predicate or without a change of quantifier. For instance, the judgment "a triangle can be right-angled" cannot be changed into "a triangle is right-angled" but only to "some triangles are right-angled" or "a right-angled triangle is right-angled". The requirement for this judgment to be real synthesis is that the subject *triangle* can be thought without the predicate *right-angled* and that other predicates (or determinants) such as *acute-angled* are possible. Therefore, it is al-

---

136 Maimon's elaborated version of this question has not been addressed by commentators, cf. Freudenthal 2006; Bergman 1967; Bransen 1991; Buzaglo 2002. I myself did not address it in my Master thesis entitled "Salomon Maimon's Concept of Number and His Criticism of Kant's Synthetic *a priori* Judgements" (Tel Aviv University, 2011), which dealt with this question in relation to arithmetic.

137 According to Maimon, if *b* is a possible determinant of *a*, a new object *ab* arises (e.g. the judgment: "a triangle can be right angled"). If *b* is a necessary determinant of *a*, a new object does not arise, but rather *a* and *b* are connected in an object (e.g. the judgment: "a triangle has three angles"). If *b* is contained in the concept *ab*, then the connection between *a* and *b* is not merely in the object, but rather in the concept itself (*Logik*, p. 50).

138 Here I follow Maimon's definitions that the judgment "*a* is *ab*" is only problematic since it is not necessary that "*a* is *ab*", as *a* can be thought without *b*. The judgment "*ax* is *a*" is apodictic since *a* is necessarily contained in *ax* (*Logik*, p. 70). In determining a connection of subject and predicate, which is possible as a problematic judgment, Maimon follows Kant (*CpR*, § A74–75/B100, p. 109–110). However, while Kant describes how some problematic judgments can be transformed into assertoric and then apodictic judgments (once the connection between subject and predicate is proven necessary; *CpR*, § A76/B101, p. 110), according to Maimon in the case of "*a* is *ab*", the problematic judgment cannot be shown to be apodictic, since the connection between *a* and *b* remains only possible and is known not to be necessary.

3.7 How are Synthetic *a priori* or Ampliative Analytic Judgments Possible? — 75

ready known that by analyzing the concept *triangle* the predicate *right-angled* would not be found. Consequently, synthetic judgments cannot be shown to be analytic in the narrow sense, in which the determinant *b* is connected with the determinable *a* in the concept (so that their relation is of objective necessity).

(b) Ampliative analytic judgments, such as "a triangle has three angles": the relation of the determinant to the determinable is of necessity. The determinant *b* is a necessary determinant of *a*, and the two are connected in the object.[139] This kind of propositions is considered by Maimon at times as apodictic and at times as assertoric, but not problematic.[140] For instance, In *Logic* he refers to the proposition "A three-sided figure has three angles" as apodictic (*Logik*, p. 405). At the least, we cannot determine that the predicate *three angles* is not contained in the concept of the subject *triangle*. In other words, unlike in the case of synthesis, the possibility of this predicate to be contained in the concept of the subject has not been disproved. It is possible to show that *predicatum inest subjecto*.

The inclusion of the ampliative notion of analysis in Kant's question is based on Maimon's new notion of analysis as grounded not only on the principle of contradiction, but also on intuition. As long as he held that there is only one kind of analytic judgments, he addressed this question only in relation to synthetic *a priori* judgments. In *Essay on Transcendental Philosophy*, for example, he refers to analytic propositions as propositions that are based on the principle of contradiction alone and refer to the object in general. Synthetic propositions are defined as grounded in intuition as well as in the principle of contradiction and refer to determined objects (*Tr.*, p. 172). There, he argues that propositions that might seem to our limited finite understanding as synthetic *a priori* can poten-

---

[139] According to Freudenthal, Maimon was influenced by the Aristotelian notion of *idion* (*proprium* in Latin, or *Segula* in Hebrew) when working on propositions where a predicate is 'coextensive with the "essence" of the substance but not included in its definition' (Freudenthal 2006, p. 4).

[140] At times, Maimon describes propositions where the predicate is not contained in the concept of the object but rather in the object itself as assertoric and not apodictic judgments (*Logik*, p. 70). However, when claiming that these assertoric judgments are incorrectly called apodictic, he mentions that they are not apodictic judgments in the narrow sense (judgments in which the predicate is contained in the concept of the subject), thus leaving room for a broader definition of apodictic: a necessary connection between a subject and predicate in the object and not in the concept. In *Logic*, Maimon states that in mathematics there are no assertoric judgments (*Logik*, p. 429), whereas in his first book he maintains that synthetic propositions of mathematics are assertoric judgments (*Tr.*, p. 185). In either case, the connection is not problematic.

tially be proven to be apodictic (if we can show the inner ground for connecting subject and predicate). To the infinite understanding they are analytic. That is to say, there might be propositions that seem synthetic *a priori* to the finite understanding but it does not mean that it is impossible to prove that these propositions are in fact analytic (*Tr.*, p. 61–62). However, once Maimon presented his new definitions of analysis of the object, he thought it fit to address Kant's question to ampliative analytic propositions as well. Although Maimon himself does not relate to this matter, it can be inferred that his argument in *Essay on Transcendental Philosophy* is valid only for ampliative analytic propositions, not for synthetic ones. In other words, synthetic propositions (in the narrow sense) cannot be reduced to identity judgments (i.e., analytic judgements in the narrow sense).

As long as one refers to synthetic *a priori* judgments in the Kantian sense, as one kind of judgments, one can always hold Maimon's initial claim that synthetic *a priori* judgments can be shown to be analytic (in the narrow sense). But once Maimon's definitions of real synthesis and of ampliative analysis are applied, it is impossible to demonstrate that synthetic *a priori* judgments (in the narrow sense) are analytic (in the narrow sense) without altering the proposition. This is true for the finite understanding. As for the infinite understanding, this matter will be addressed in the end of the next section.

3.8 Analysis and Discovery, Synthesis and Invention

In concluding this chapter, I would like to dedicate a few words to the relations between discovery, invention, analysis and synthesis. This chapter reviewed the different notions of invention (used here as a general term that includes both invention and discovery) as different kinds of analysis and synthesis. Comparing Maimon's definitions, one finds that analysis in its broader sense and discovery share the same definition and so do synthesis and invention (in the narrow sense).

Maimon himself does not make this connection between these four concepts. He mentions however, that both analytic and synthetic thought extend our knowledge. The only difference is that analytic thought extends our knowledge by introducing a new determination of an already thought object whereas synthetic thought extends our knowledge by introducing a new object (*Logik*, p. 29–30). This differentiation is equivalent to the difference drawn by him between discovery and invention: to discover is to attribute a new property to a given object and to invent is to present a new object (*Ueber den Gebrauch*, p. 8; *Das Genie*, p. 364). Consequently, although not explicitly mentioned by him, in Maimon's view analysis of the object is discovery and synthesis is invention. Since Maimon treats the difference between discovery and invention as subjective, the question arises whether the difference between ampliative analysis and synthesis is subjective. If the answer is proven negative, and the difference between discovery and invention in Maimon's theory is a real difference, then his presentation of the theory and methods of invention is somehow lacking. If that were the case, he should have presented a theory of invention and a theory of discovery, each accompanied by its own methods.

Let us first examine the claim that the difference between discovery and invention is only subjective. Maimon claims that in the broader sense they both have the same meaning: "to bring out unknown truths following certain methods" (*Das Genie*, p. 363; *Ueber den Gebrauch*, p. 10), thus presenting a more general shared definition. He states that his theory deals with invention in the broader sense, which includes both invention and discovery (*Das Genie*, p. 363; *Ueber den Gebrauch*, p. 10). Once he had marked the difference between discovery and invention, he did not mention their difference any further. In *Methods of Invention*, the difference between discovery and invention is not even mentioned. Another important aspect is that his methods are not devoted separately to discovery or invention but meant to be used for both processes. Were that not the case, his last work on invention would have been called *Methods of Invention and of Discovery*. Furthermore, the same method may be used as a method of invention and as a method of discovery. For instance, Maimon's second kind of analysis ("analysis of a complex problem or a complex proposition into a simple one") can be used in *Elements* I.1 when constructing an equilateral triangle and also in proposition 39 of Euclid's *Data* to prove that if each of the sides of a triangle are given in magnitude, then the triangle is given in form (see Section 4.1.2). At times Maimon interchanges between the verbs *to discover* and *to invent:* when in *Logic* he refers to how fruitful each kind of judgment is, he uses the verb *to discover* [*entdecken*] to describe inventing a new object (*Logik*, p. 53). Another example of interchanging the two terms is found in his preface to the translation of Bacon's *Novum Organum:* he mentions that he presents "the important mathematical discoveries" [*mathematischen Entdeckungen*] (*Baco*, p. CXII), whereas the title of the work is *A Short Exposition of Mathematical Inventions*. Rather than explaining this as Maimon being inconsistent with the use of his own definitions, it seems that it is due to the lack of objective difference between inventing and discovering.

Let us now consider as true the assumption that invention and discovery differ only subjectively. If we take this assumption together with Maimon's definitions of analysis of the object and synthesis, then following this proportion (if I may borrow a mathematical term) it can be inferred that the difference between analysis of the object and synthesis is subjective. To support this claim, I propose the following argument: as mentioned earlier, the inventive faculty is considered by Maimon as that which separates man from other animals and makes him similar to God (*Ueber den Gebrauch*, p. 5). In *Giv'at Hamore* he describes man's creation in the image [צלם] of God as forming concepts and deriving attributes from them, noting that the difference between finite and infinite understanding is only in degree. The difference between the two is only subjective: the finite understanding either forms concepts and then arrives at showing that they are real; or it first cognizes a given object in actuality (or in experience) and then forms a concept. The infinite understanding, however, thinks concepts in such a way that by the thought of their concepts, real objects are determined simultaneously. The activity of thinking by the infinite understanding is not within the form of time, the concepts are not thought before or after the presentation of real objects (my use of the word *simultaneously* is due to shortcomings of language, and is not meant literally as "at the same time"; *Giv'at Hamore*, p. 33–34).[141] Maimon's example of the number 2 as ratio is an example of how the understanding can produce pure concepts without real objects being given. It determines real objects with the creation of the concepts. The number 2 is created as a real object at the same time its relation-concept [*Verhältnisbegriffe*], the ratio 2:1, is thought (*Tr.*, p. 190). What seems to us as synthetic knowledge is analytic for the infinite understanding since it "thinks all possible objects according to an analytic rule; from this it follows that the forms or synthetic rules have objective necessity only for us (since we cannot make them analytic on account of our limitation), but not in themselves" (*Tr.*, p. 418). Another manner with which to approach the difference between the finite and infinite understanding is by using the terminology *given* and *sought*. According to Maimon, in real synthetic thought the determinable is given and the determinant is sought (together creating a new determination). In real analytic thought, the determination (including both determinable

---

[141] A similar claim appears in *Prg*, p. 20, where he writes that the finite understanding thinks all mathematical concepts at the same time [*zugleich*] the cognition exhibits them via construction *a priori* as real objects, and that in this we are similar to God.

Freudenthal mentions that Maimon's adoption of Maimonides' doctrine of the unity of knowledge, the knower and the known lead him to the Leibnizian view of intuition as unclear thought, i.e. that intuition and thought are entities of the same kind and thus do away with the *quid juris* question (Freudenthal 2011, p. 129).

On the difference between Maimon's notion of the difference between finite understanding and infinite understanding as only formal and between Kant's view that this difference is real, see Franks, 2003.

and determinant) is given and the determinant is sought (*Logik*, p. 28).[142] This is true for the finite understanding. For the infinite understanding, on the other hand, there is no given and no sought, only thought. Consequently, we can claim that what seems to be a real difference for the finite understanding is not a difference for the infinite understanding. Whereas the finite understanding is unable to think all possible determinants of a determinable as necessary in one determination (so that, for instance, a triangle is either right-angled or obtuse-angled, but is not both), the infinite understanding thinks all possible determinants as necessary determinants of a determinable all at once.

Since for the infinite understanding there is no such difference and since our understanding is its schema, the difference between arriving at a new attribute or at a new object is only subjective. Invention and discovery, synthesis and analysis, are all names of methods used by the finite understanding to describe its different modes of action. However, they are one and the same for the infinite understanding, whose action is of constant thinking and creation, even if this infinite understanding has only the status of an idea in Maimon's philosophy.

---

142 Maimon writes that in analytic though the sought is the determinable [*Bestimmbare*]. However, it should be the *determinant* that is sought. For instance, if we are given the determination *equilateral triangle*, we would look for the determinant *equiangular* and not for the determinable *triangle*. Were his intentions to say that in the determination *equilateral triangle* we wish to arrive at *triangle*, then his use of the terminology *sought* [*wird gesucht*] is incorrect. Moreover, when presenting an example, Maimon suggests looking for the determinant and not the determinable. He writes that in the judgment: "triangle (space enclosed by three lines) has three angles", *three angles* is not contained in the concept of *triangle* but is educed from the subject *triangle* itself (that is, the determinant *three angles* is educed from the object and not the concept of triangle; *Logik*, p. 29). According to this example, the given is the determination and the sought is the determinant *three angles*. It is not mentioned that the determinable *triangle* is sought.

Chapter 4: Methods of Invention

Maimon's methods of invention are designed to be used by both mathematicians and philosophers. Although his articles on invention were published in philosophical journals, he addresses his work to mathematicians as well, suggesting that professional mathematicians could join him in developing his work further:

> I would be very pleased if professional mathematicians who do not practice their science mechanically but reflect on it, awaken by these suggestions, would join me in the realization of such an important plan (since I am only an amateur in this science; *Das Genie*, p. 383–384).

His aim to advance not only philosophical, but also mathematical knowledge is explicitly expressed, naming the type of mathematicians that would take interest in his work as "philosophical mathematicians":

> I wish that the philosophical mathematicians acknowledge the extent and importance of such an enterprise, and would like to join me in its execution. *(Ueber den Gebrauch*, p. 35)

His address to the philosophical mathematicians, rather than to philosophers who are interested in mathematics, emphasizes that his project has more than philosophical aspirations, aiming at the advancement of mathematical knowledge. The small scale of this advancement corresponds to his idea of the methodical inventor as one who advances "step by step" rather than in giant leaps. Moreover, his appeal to mathematicians corresponds to his claim that philosophy can offer only formal inventions and "only pure and applied mathematics can rightly claim to make inventions" (*Das Genie*, p. 372). His appeal to mathematicians may not have reached as far as he wished since there is no evidence that his methods were indeed applied by mathematicians. Nevertheless, addressing this group of scholars (together with philosophers) accords with his endeavor to contribute to the advancement of knowledge. His intellectual approach to knowledge wherein truths are connected to each other has a specific expression in his remark that philosophical truths and mathematical inventions should be examined together, in order to shine a brighter light on each of these fields (*Baco*, p. CXII). Following Maimon's statement that "everyone will be able to test whether he is able to *invent* something new by these *methods*" (*Ueber den Gebrauch*, p. 4), this chapter includes new examples to which his methods could be applied.

Maimon's contribution to the expansion of mathematical knowledge lies neither in presenting many new proofs and solutions nor in introducing a new mathematical field, but rather in offering a new perspective on approaching propositions and problems. He himself states that he extracts his methods from actual inventions (*Ueber den Gebrauch*, p. 17). In most cases Maimon did not invent a completely new method but rather improved or better articulated a given method of invention. An example of an improvement of a mathematical practice into an explicit rule is the fourth kind of analysis: analysis of the object. The alteration of a given object in order to arrive at the sought solution was a known practice, yet it was Maimon who presented it as a method. An example of a new articulation of a known rule of invention is the first method: analysis of the conditions of a problem or proposition into true and pseudo-conditions. While the idea that we should examine whether a given condition of a problem is necessary had already been mentioned by Descartes (*CSM*, Vol. 1, p. 436), Maimon's new articulation of the rule using the terms *true conditions* and *pseudo-conditions*, followed by his example, has proved productive when searching for pseudo-conditions in Euclid's *Elements*. Even though Maimon's methods are not completely new, they do allow us to study Euclid's *Elements* and *Data* anew and to present new proofs and solutions accordingly.

The methods of invention presented by Maimon are accompanied by geometrical examples, all from Euclid's *Elements*. More accurately, almost all of his examples are from Book I of *Elements*, alongside one example from Book III. It is unclear which version of *Elements* Maimon used. He cites Johann Christoph Schwab's German translation of Euclid's *Data* (published in 1780) which also includes a translation of the first six books of *Elements*.[143] Schwab's translation of *Elements* is based on Robert Simson's English version (first published in 1756).[144] Another possible version of *Elements* that Maimon could have used is Lorenz's German translation, published in 1773 (Books I–VI) and 1781 (Books XI–XII) (Heath 1956, Vol. I, p. 107).[145] Maimon mentions Clavius' commentary on *Elements* I.16 (*Ueber den Gebrauch*, p. 32–33; *Das Genie*, p. 382), so at least at

---

143 Schwab, Johann Christoph, *Euklids Data, verbessert und vermehrt von Robert Simson, aus dem Englischen übersetzt und mit einer Sammlung Geometrischer, nach der Analytischen Methode der Alten aufgelößter Probleme begleitet*, Stuttgart: Christoph Freiderich Cotta, 1780. It is a translation from Simson's English translation of the *Data* (1762). Simson's translation of *Elements* was published in 1756 and only in 1762 was *Data* added.
144 Simson's version includes his own proofs for some of the propositions as declared in the title: *The Elements of Euclid, vis. The first six Books together with the eleventh and twelfth. In this Edition the Errors by which Theon and others have long ago vitiated these books are corrected and some of Euclid's Demonstrations are restored.*
145 According to Heath, this translation was used in schools and was "the first attempt to reproduce Euclid in German word for word" (Heath, 1956, Vol. I, p. 107).

one point in his career he either read or was taught Clavius' version of Euclid's work (first published in 1574 in Latin).[146]

The similarity between Maimon's methods and methods mentioned in Proclus' commentary on Book I of *Elements* has led me to the realization that Proclus' commentary had a great influence on Maimon.[147] It is likely that Maimon was acquainted with Proclus' commentary via Clavius' work.[148] A second plausible option is that Maimon had access to Taylor's English translation of Proclus' commentary on Euclid's *Elements*, published in 1789.[149]

By extracting methods from actual inventions and refining already existing methods, Maimon's work on the methods of invention can be largely summarized by the terms *method* and *order*. This accords with the view stated in *Giv'at Hamore* that we cannot invent without order and method (*Giv'at Hamore*, p. 19).[150] If I may adopt the metaphor of craftsmanship (previously used by Bacon and Descartes), Maimon's work on the methods of invention can be

---

146 Clavius' version of the *Elements* is not a direct translation of Euclid's work. He rewrote proofs and added to them as he saw best (Heath 1956, Vol. I, p. 105).
147 In addition to the similarities between Maimon's work and Proclus', which will be discussed in this chapter, my assumption also relies on Maimon's remark that Proclus presented alternative proofs to Euclid (*Ueber den Gebrauch*, p. 31). Proclus was the only commentator mentioned by Maimon when discussing alternative proofs of the *Elements*, which may suggest that he was familiar with his commentary.
148 On the influence of Proclus on Clavius' commentary of *Elements*, see: Claessens 2009, p. 317 and Sasaki 2003, p. 50.
149 *The Philosophical and Mathematical Commentaries of Proclus, on the First Book of Euclid's Elements, to which are added A History of the Restoration of the Platonic Theology, by the Latter Platonist, Vol. II*. Thomas Taylor (Tr.), London: Payne and Son et al.
150 These terms were not only often used in the field of theories of invention, but they are also etymologically related. As mentioned by Wallace, the meaning of the Greek word *methodos* is *means*, and it is closely related to the notion of *order*. Method is a means to achieve an end, by ordering materials for the realization of that end (Wallace 1973, p. 249). Aristotle uses the term *methodos* with the term *techne* (in *Rhetoric*) and with the term *hechein* (in *Topics*), both terms meaning "to have a plan or system, to work or practice by rule" (Wallace 1973, p. 249). Similarly, in the 16th century *Method* was used both for describing procedures for discovering new knowledge and for arranging existing knowledge for communication (known as the *art of discourse*) (Jardine 1974, p. 2). This relationship is found also in the definitions of *method* presented in the *Encyclopédie*. For instance, method is said to be "the order which we follow in order to find the truth or teach it" ("Méthode [*Logique*]", *Encyclopédie*, Vol. 10, p. 445). Moreover, *order* is the first term chosen by Jaucourt to define *method*, alongside *rule* and *arrangement* (Jaucourt, "Méthode [*Arts & Sciences*]", *Encyclopédie*, Vol. 10, p. 460).

best described as sharpening the tools in the toolkit of the *ars inveniendi* tradition.[151]

4.1 Seven Kinds of Analysis

The seven kinds of analysis are the core of Maimon's methods of invention. At first, it might seem puzzling that Maimon chose to present examples from Euclid's *Elements*, a work considered to be a "book of the treasury of synthesis" rather than from *Data*, considered to be a "book of the treasury of analysis" (Hintikka & Remes 1974, p. 10).[152] However, analysis is a practice that should not be reduced to mean merely the reverse of the process of synthesis, such as described by Pappus and following him, by many philosophers and historians of mathematics.[153] Rather, as stated by Mahoney, analysis is "a body of problem solving techniques" that can also be a "problem generator" or a "mathematical toolbox" (Mahoney 1968, p. 323; p. 320).[154] As mentioned in the previous chapter, Maimon's practice of the methods of analysis is based upon the notion of analysis in its broader sense, one that resembles the practice of *diorism*.

This approach to analysis as including more than logical analysis and as a problem-solving toolbox is best expressed by the fact that Maimon's methods of analysis are better understood when accompanied by examples. For instance, it is hard to fully capture Maimon's method of analysis of an object when it is presented without an example (*Das Genie*, p. 379–380). When followed by an example (*Ueber den Gebrauch*, p. 29–31), we have a better understanding of what applying the method entails. Reading the general description of the method, it is left unclear how we should apply it: instructions such as "the object of the proposition to be proven or of the problem to be solved, must be transformed into such an object, to which the premises refer *immediately*, by increase or diminution, or another alteration" (*Das Genie*, p. 380) are not very informative as to how such a transformation should be made. Presumably, by *alteration* (increase or di-

---

[151] Bacon and Descartes used metaphors of 'instruments' or 'machinery' in order to describe the art of invention. Just as craftsmen use instruments in order to achieve greater quality in their production of objects, so logic should develop its own tools in order to advance discoveries and inventions (Buchenau 2013, p. 22–24).
[152] The division into "books of the treasury of analysis" and "books of the treasury of synthesis" is Pappus'. Marinos, a known commentator of *Data*, makes a similar claim stating that in *Data* Euclid establishes the basis for the analytic method (Marinos 1977, p. 105).
[153] E.g. Robinson 1936, p. 472.
[154] Among the techniques mentioned by Mahoney are reduction and neusis (Mahoney 1968).

minution) Maimon refers to magnitudes given in the problem and to actions such as prolonging a given line. However, once the example is added, the reader has a better notion of what this analysis consists of: in order to prove the equality of angles in one isosceles triangle given in *Elements* I.5 (triangle *ABC*), we should use *Elements* I.4 which demonstrates equality of angles in equal triangles. Since *Elements* I.4 proves the equality of two angles that are enclosed in two different equal triangles (each an angle in one triangle) and since the angles in *Elements* I.5 are enclosed in the same triangle (the angles at the base of an isosceles triangle or their adjacent angles), the triangle given in *Elements* I.5 needs to be altered. This change includes extending the equal sides of the triangle and constructing two other lines. After this alteration of the given triangle in *Elements* I.5 we receive new triangles (triangles *AFC*, *AGB*, *BCF* and *BCG*). To these triangles, we can apply *Elements* I.4 and show equality of triangles and then equality of angles. Consequently, we can demonstrate the equality of the angles under the base of the original triangle and thus also the equality of the angles at the base of the original triangle. Only when presented with a specific example can one understand what "transforming a given object in such a way that we could apply another proposition to it" means, in a way that is possible to apply this rule to a given problem or proposition.

This approach to methods of analysis as a toolbox is closely related not only to Greek mathematical practices, but also to ancient practices in rhetoric. As mentioned by Bernard, practice was at the core of ancient rhetoric and was intended to prepare the rhetorician for dealing with all possible cases in advance (Bernard 2003b, p. 409). More specifically, *loci communes* served as tools for the invention of arguments and were used only in particular cases, so that one could not simply learn the general rules but had to learn them through practice (Leff 1996, p. 449). Because Euclidean geometry is closely related to ancient Greek rhetoric and Maimon's own theory of invention is closely related to the tradition of the art of arguments, it is comprehensible that Maimon's methods of analysis are practical tools to be applied in specific cases more than they are general rules.

4.1.1 Analysis of the Conditions of a Problem or Proposition

The first kind of analysis is the analysis of conditions. Maimon mentions three types of conditions, referring to different aspects of the problem or its solution (or different aspects of the proposition or proof). Each type of condition will therefore be presented and discussed separately.

4.1.1.1 True and Pseudo-Conditions

Maimon's analysis of conditions is aimed at distinguishing true conditions from pseudo-conditions. According to Maimon, a true condition is a necessary condition. A pseudo-condition is mistakenly taken for a condition, when one can solve the problem without taking this condition into account. He presents *Elements* I.1 as an example of a problem that includes both true and pseudo-conditions: "On a given finite straight line to construct an equilateral triangle".[155] The triangle and the equilateral, writes Maimon, are both true conditions of what is sought in the problem. While these are necessary conditions, the "givenness" [*Gegebensein*] of the given line *AB* is a pseudo-condition since the line can be taken arbitrarily. The determination of the line does not change the solution to the problem. Maimon therefore suggests that we can rephrase the problem as "to construct an equilateral triangle" (*Ueber den Gebrauch*, p. 24–25). It is possible that he was influenced by a similar distinction between true and pseudo-conditions made by Descartes in his *Regulae*.[156] Descartes states that we should consider whether a condition is necessary when defining a problem (*CSM*, Vol. 1, p. 436).[157] This distinction is an important part of our toolkit for invention, as it allows us to isolate the conditions and better articulate which ones we need to

---

[155] Maimon presents the problem as the particular instance of the given line *AB*, and not as a general statement about a given finite straight line, as usually appears in editions of the *Elements* (for example: "To describe an equilateral triangle upon a given finite straight line"; Simson 1811, p. 16; and "On a given finite straight line to construct an equilateral triangle"; Heath 1956, Vol. I, p. 241). This particular instance, however, reproduces the conditions of the general statement: that the line is straight and finite.

[156] The issue of falsely perceiving something as true already concerned Maimon when he wrote his philosophical dictionary in 1791. In the entry "Deception and Pretense" [*Täuschung und Schein*], deception is defined as the confusion of a representation with the object, or the assignment of a judgment about the object to its representation. This kind of confusion occurs when we look at a painting and forget it is a painting, but regard it as the natural object, or when we watch an actor and assign his actions to the character he is playing. Pretense is the confusion of the object with its representation, or the assignment of a judgment about the representation to the object. This kind of confusion takes place when we watch the sun setting over the horizon: we confuse the position of its image over the horizon (which we glance at) with the sun's true position, below the horizon. We can know its true position by using the optical law of refraction (*PhW*, p. 121–122). Deception and pretense act in such a way that, even after we know we were deceived, the deception continues to affect us. For instance, even after we know the true position of the sun (below the horizon), we will still see its image in the position where we formerly believed the sun to be (over the horizon; *PhW*, p. 125).

[157] He does not use the term *pseudo-condition* but refers to conditions that only appear to be conditions (*CSM*, Vol. 1, p. 438).

fulfill in order to find what is sought.¹⁵⁸ For instance, in Maimon's example of *Elements* I.1 he correctly identifies triangularity and equilaterality as the true conditions of the problem. However, his example is not exhaustive. Whereas Maimon mentions *triangle* and *equilateral* as necessary conditions, he fails to mention that a straight line must be given. When he refers to the "givenness" of the straight line as arbitrary, he means that the magnitude of the line can be determined arbitrarily. However, the condition stating that "a straight line is given" is necessary and not a pseudo-condition. Moreover, we should break this "givenness" down: in the case of a straight line, being given in magnitude and being given in position are necessary conditions.¹⁵⁹ Where it is posited and the determined size of the line may be taken arbitrarily, yet it is necessary that the straight line is given in magnitude and in position.¹⁶⁰

Let us consider another example of a pseudo-condition. In *Elements* I.12, "To a given infinite straight line, from a given point which is not on it, to draw a perpendicular straight line" (Heath, 1956, Vol. I, p. 270), the condition "infinite" is a pseudo-condition.¹⁶¹ This proposition is presented by Maimon as an example of a condition of the possibilities of a solution (a method that will be discussed in the following section). Maimon mentions that the condition *infinite* is a condition of the possibility of a solution in certain cases, even if not in all cases, but he does not elaborate further (*Ueber den Gebrauch*, p. 25). Presumably, this condition is meant to ensure that the circle cuts the given line twice. I propose that the condition *infinite* is not a necessary condition, neither of the problem nor of its

---

158 This method can be used in other fields of knowledge. Hintikka, for example, points out that even though Kant claims that sensibility is a condition for representing objects in intuition (which is a particular representation), it is not a condition. Only individuality is the condition (Hintikka 1992, p. 23). Hintikka of course did not intentionally use Maimon's method, but his argument is an example of a possible application of such a method.

159 Def. 1: "'Given in magnitude' is said of areas and lines and angles to which we can get equals" (Taisbak 1991, p. 147); Def. 4: "'Given in position' is said of points and lines and angles which always occupy the same place" (Taisbak 1991, p. 142–143). I adopt Euclid's viewpoint that a straight line is not considered as *figure* (Heath 1956, Vol. I, p. 182–183). Therefore, unlike a rectilineal figure, it is not said to be given in form (Def. 3: "A rectilineal figure is given in form if and only if each of its angles is given and the ratios of its sides to each other are given"; Taisbak 1991, p. 154).

160 Once the positions of the extremities of the line are determined, so is the magnitude of the line determined (*Dt*. 26: "If the extremities of a straight line be given in position, the line is given in position and in magnitude"; Taisbak 2003, p. 99).

161 This condition appears in other translations of *Elements* as well. For instance, Lorenz uses the word "infinite" [*unbegränzte*] (Lorenz 1824, p. 10). In Simson's translation, the condition appears as "unlimited": "To draw a straight line perpendicular to a given straight line of an unlimited length, from a given point on it" (Simson 1811, p. 25).

solution, and I offer a case in which the given problem can be solved without this condition. The proof of this proposition begins with the construction of a circle in which the given point C (which is not on the given line) is the center of the circle. The circle meets the given infinite line AB in two points, G and E (this circle is constructed using a random point D, which is on the other side of the line AB). The line GE is then bisected by the point H; the proof goes on to demonstrate that GH is equal to HE, with HC as a common side, and since CG equals CE (both are radii), it is demonstrated that the two triangles HGC and HEC are equal. Therefore, the angle CHG is equal to the angle EHC. Angles CHG and EHC are adjacent and equal, so that they are each a right angle. Therefore, the line CH is perpendicular to AB (Heath 1956, Vol. I, p. 270–271).

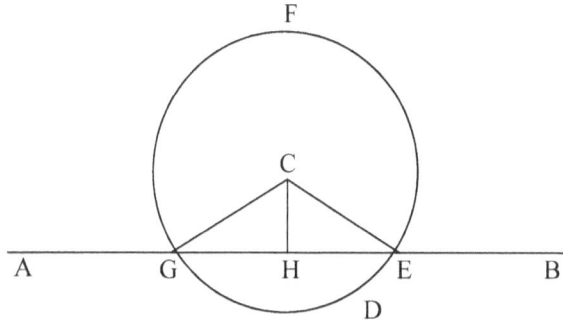

Fig. 1. Based on a diagram from Heath 1956, Vol. I, p. 271.

In the solution above, the true condition is that the line AB needs to be more than twice the line HE so that we are ensured of the ability to cut the given line AB twice. There is no requirement for the line AB to be infinite in order to be able to construct a circle, only that it is long enough for such construction. The proof is not dependent on the property of being infinite but only on the condition that a circle could be constructed on the line AB so that the given point C is its center and so that the given line AB is cut twice by the circle. The problem can be solved with a finite line, not only with an infinite line. Therefore, *infinite* is not a necessary condition.

A closer look into the problem and solution shows that the condition that the line AB will be cut twice by the constructed circle is a condition solely of the solution proposed by Euclid. It is not a true condition of the problem since the same problem presented in *Elements* I.12 can be solved as follows: Let AB be the given straight line and C the given point which is not on it. It is required to draw a perpendicular straight line from the point C to the straight line AB. Let a point H be taken at random on the given straight line AB. Let the circle DGH be described so that the given point C is its center and so that the line AB touches the circle at the point H, using the radius CH.

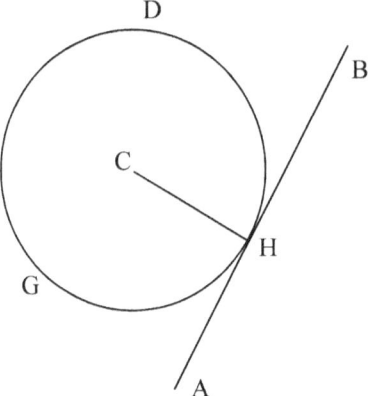

Fig. 2.

At this point, the problem presented in *Elements* I.12 is similar to the problem in *Elements* III.18: "If a straight line touch a circle, and a straight line be joined from the center to the point of contact, the straight line so joined will be perpendicular to the tangent"; Heath 1956, Vol. II, p. 44). We can construct a perpendicular to the tangent of a circle without the condition that the tangent is infinite.[162] It should be mentioned that *Elements* III.18 is proven without using *Elements* I.12 at any stage of the proof: Euclid's proof of *Elements* III.18 is based on *Elements* I.17 and I.19. These propositions are proven based on *Elements* I.5, I.13, I.16 and I.18. The latter four do not require the use of *Elements* I.12 for their proofs, either directly or indirectly. We can therefore offer a solution to *Elements* I.12 using a finite line and not an infinite one. Consequently, the condition that the given straight line AB is infinite in *Elements* I.12 is a pseudo-condition.[163]

---

[162] Infinite length is not mentioned in the definition of a tangent, in *Elements*, Book III, Def. 2: "A straight line is said to touch a circle which, meeting the circle and being produced, does not cut the circle" (Heath 1956, Vol. II, p. 1)

[163] Heath writes in his commentary that "the circle described with CD as radius must necessarily cut AB in two points" (Heath 1956, Vol. I, p. 272). He also adds that Proclus refuted the suggestion that the circle cuts the line AB in three or more points (Heath 1956, Vol. I, p. 272). Within the framework of this solution, these statements are true. But the condition that the circle will cut the given line at two points is not necessary to the problem presented, only to the solution that is offered.

4.1.1.2 Conditions of Possibility of the Solution

Another type of condition mentioned by Maimon is the conditions of possibility of the solution. As mentioned in the previous section, he presents *Elements* I.12 as an example, claiming that *infinite* is a condition of the solution (*Ueber den Gebrauch*, p. 25). Maimon does not elaborate much on this method or its example, yet it can be assumed that examining what enables the solution requires verifying whether a condition is necessary or not. Heath's commentary to *Elements* II.14 provides an example of how the method of distinguishing between true and pseudo-conditions can be applied to the method of finding the conditions of possibility of the solution. In *Elements* II.14, "To construct a square equal to a given rectilineal figure" (Heath 1956, Vol. I, p. 409), we are required to construct a parallelogram *BD* equal to the given rectilineal figure *A*. Euclid first presents the case in which the sides *BE* and *ED* of parallelogram *BD* are equal, so that the parallelogram is a square. Then presenting the case in which one of the sides of the parallelogram is greater than the other he writes "Let *BE* be greater" (Heath 1956, Vol. I, p. 409). Heath quotes Simson's commentary to the effect that "it was not necessary to put in the words "Let *BE* be greater", since the construction is not affected by the question whether *BE* or *ED* is the greater" (Heath 1956, Vol. I, p. 410). Although Simson is correct to mention that it is not necessary for *BE* to be greater than *ED* and vice versa, the problem can be solved with neither *BE* nor *ED* being greater. Hence, the condition of having one side of the parallelogram be greater than the other is a pseudo-condition.

4.1.1.3 Conditions of the Problem

Maimon presents an example of the condition of the possibility of the problem in *Elements* I.22: "Out of three lines, which are equal to three given straight lines, to construct a triangle: thus it is necessary that two of the straight lines taken together in any manner should be greater than the remaining one" (Heath 1956, Vol. I, p. 292). He mentions "the two straight lines taken together need to be greater than the remaining line" as a necessary condition of the possibility of the problem for constructing a triangle (*Ueber den Gebrauch*, p. 25). In this case, the condition is demonstrated by Euclid in *Elements* I.20: "In any triangle two sides taken together in any manner are greater than the remaining one" (Heath 1956, Vol. I, p. 286). Conditions of the problem can be either the conditions of what is given or of what is sought. In order to explain these two types of conditions, Maimon presents *Elements* I.2: "From a given point to draw a straight line equal to a given straight line."[164] The solution of the problem presented in the *Elements* uses a point positioned outside the given line. Maimon mentions two other possible conditions of the given point: the first is when the given point is positioned at one end of the given line. This condition facilitates the proof. The second possibility is that the given point is positioned on the given line (but not at one of its ends). This makes the proof much more difficult. The first option regards the "givenness" of the point as part of the problem: the person solving the problem reads the condition "at a given point" as the condition "at an arbitrary (but not at every) point". Consequently, writes Maimon, we can choose the position of the point to be at one of the endpoints of the line and therefore solve the problem more easily. This is the approach of the person solving the problem. The second option regards the "givenness" of the point as a required condition of the problem: one has to read the condition "at any given point" as meaning "at each given point". Maimon comments that this is the approach of the person presenting the problem (*Ueber den Gebrauch*, p. 25–26). Maimon's introduction of the two additional cases of the problem is accompanied by an implicit solution for the case where the given point is positioned at one end of the given line. The implicit solution is to construct a circle in which the given line serves as the radius and the given point is the circle's center point. This solution is trivial since any other radius of the circle will solve the problem.[165] To the second case added by Maimon, where the given point is positioned on the given line (but not at one of its ends), he does not offer a solution (the solution here is less trivial). These additional cases appear already in Proclus' commentary on Euclid's *Elements*. Proclus suggests that the given point is either on the given straight line or external to it. If it is on the given line,

---

[164] Here I am using Simson's translation of *Elements* I.2 (Simson 1811, p. 16). Maimon's phrasing is unclear ("To place an equal line at a given point of a given line"; *Ueber den Gebrauch*, p. 25). Heath's translation already entails a condition of the problem: "To place at a given point (as an extremity) a straight line equal to a given straight line" (Heath 1956, Vol. I, p. 244). He explains his addition as follows: "(as an extremity). I have inserted these words because "to place a straight line *at* a given point" [...] is not quite clear enough, at least in English" (Heath 1956, Vol. I, p. 244). Heath's addition "as an extremity" could be misinterpreted to mean that the given point (*A*) is an extremity of the given straight line (*BC*). The diagram of the solution shows that the given point *A* is not on the given straight line *BC*, and that by "as an extremity", Heath means that the constructed line (*AL* – equal to the given line *BC*) will be placed such that the point *A* will be one of its extremities.

[165] Heath mentions that the reason Euclid presents only one case is that it was uncommon for ancient Greek geometers to present several cases. Thus, Euclid presented only the most difficult case, assuming the reader would be able to arrive at the solutions of simpler cases himself (Heath 1956, Vol. I, p. 246).

then it can be either at one of its extremities or between them. If the point is external to the given line, then it has a lateral position that either forms an angle with it or is in a direct position (Proclus 1789, p. 34). Proclus raises the questions of what happens if the given point is given as being at one extremity of the given line and if the given line is infinite. He wonders why Euclid did not specifically declare the condition that the given straight line is finite and not infinite, especially since Euclid does at times make this distinction. He presumes that Euclid must have assumed that the given straight line is finite but did not mention it as a condition in the problem (Proclus 1789, p. 33–34). It is very plausible that Maimon was directly influenced by Proclus' suggestions of the additional cases and of verifying the different types of conditions.

Let us consider another example of the application of this method, *Elements* IV.1: "Into a given circle to fit a straight line equal to a given straight line which is not greater than the diameter of the circle" (Heath 1956, Vol. II, p. 80).

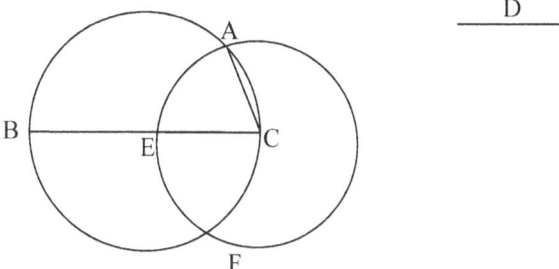

Fig. 3. Based on a diagram from Heath 1956, Vol. II, p. 80.

Circle *ABC* is given, and *D* is a given straight line (not greater than the diameter of the circle). *E* is the center of the circle *ABC*. Euclid mentions two cases. The first case is if *BC* equals the given line *D*. In this case, we arrive at the sought result very easily since *BC* is fitted into the circle and equals the given line *D*. The second case is if *BC* is greater than *D*. In this case, we construct a circle *EAF*, with *C* as its center. We take the line *CE* (which is a part of line *BC*) to be equal to the given straight line *D*. Both *AC* and *CE* are radii of the constructed circle *EAF*. Thus, the line *AC* (which equals *CE*) equals the given line *D* and is not greater than the diameter of circle *ABC* (Heath 1956, Vol. II, p. 80). Heath remarks that there is an infinite number of solutions for this problem, among them are these two difficult cases: "(1) that the chord is to be parallel to a given straight line, or (2) that the chord, produced if necessary, shall pass through a given point" (Heath 1956, Vol. II, p. 81). The first case was solved by Pappus and the second discussed by Apollonius (Heath 1956, Vol. II, p. 81). These two additional cases are an example of how considering the conditions of a problem or a proposition can lead mathematicians to new solutions and proofs, even if their inventors did not consciously use this method.

4.1.2 Analysis of a Complex Problem or Complex Proposition into Simple Ones

The second kind of analysis is possible only after the application of the first kind of analysis since, Maimon argues, one uses the conditions in order to simplify the problem or proposition. He states that mathematical problems and propositions are usually presented as simple, even when they are in fact complex. It is necessary for the invention of a solution of a problem (or a proof of a proposition) that the given complex problem be solved as a set of simple problems. For the reason that complex propositions regarded as simple can only be solved by "a lucky idea". However, if we want the solution to "be found methodically", the problem should be broken down into simple problems (*Ueber den Gebrauch*, p. 26–27; *Das Genie*, p. 378–379). In many cases, writes Maimon, solutions to problems presented in textbooks are designed for learning the outcomes, not for learning the ways in which one arrives at these solutions. He himself would rather present a longer and more elaborate solution that will reveal how the solution came to be invented. His attention is set on the art of invention rather than that of learning. He calls this method "the way of invention", adding that it is necessary for the invention of a solution or a proof (*Ueber den Gebrauch*, p. 27–28; *Das Genie:* 374). This method is not original to Maimon – it appears in Proclus' commentary on the *Elements*. Proclus differentiates between simple and composite theorems. He defines simple theorems as those having one datum and one object of investigation, such as "every isosceles triangle has the angles at the base equal" (Proclus 1789, p. 49). Composite theorems are defined as theorems composed of many particulars, which are hypotheses or conclusions. He then differentiates between composite theorems that are complex and incomplex (according to whether they can be divided into simple theorems) as well as between composite conclusions and composite hypotheses (Proclus 1789, p. 49). He suggests a method similar to the one proposed by Maimon: "The incomplex are such composites as cannot be divided into simple theorems [...] But the complex are such as may be divided into things simple" (Proclus 1789, p. 49). Whereas Proclus suggests breaking down complex propositions into simpler ones, Maimon adds that this can be done with problems as well, not only theorems. Another known proponent of such a method is Descartes.[166] Maimon's added value to this method lies in his claim that using this method can uncover the manner in which the inventor initially arrived at the proof or solution.

Proposition I.1 in *Elements* is introduced by Maimon as an example of a complex proposition presented as a simple one. While the proposition seems to be simple, phrased as "to construct an equilateral triangle on a given finite straight line", Maimon breaks it down into three conditions: 1) to find two lines that are equal to a given line and therefore to each other; 2) to connect the two found lines with the given line at both of its endpoints; 3) to connect the two found lines with each other at their other endpoints, so that an equilateral triangle is constructed. The solution of the complex problem is arrived at by solving these three simple problems. Maimon suggests solving the first problem by taking the given straight line and describing a circle with it. Then from the center of the circle, which is one endpoint of the given line, we take two radii of this circle. Hence, we find two lines that are equal to the given straight line. The second problem is solved by describing another circle from the other endpoint of the given line. Thus, each arbitrary radius of the second circle is equal to the given line and is attached to one of its endpoints. In the third problem, we take a radius from each circle. However, they cannot be taken arbitrarily. Their meeting point has to be at the endpoint of each of them (that is, they will not intersect: only their endpoints will meet). Therefore, concludes Maimon, the two lines should be drawn according to the point of intersection of the two circles. He refers to this elaborate way of solving the problem as "the way of invention" (*Ueber den Gebrauch*, p. 27–28). Maimon regards problems as conditions to be met. For this reason, when he refers to the third problem, he also names it "the condition of the presented problem" (*Ueber den Gebrauch*, p. 28). This explains his claim that the analysis of conditions must precede the breaking down of the complex problem or proposition, an idea that also appeared in Descartes' work on invention.[167] Equating the simple problems with conditions is justified since we divide the complex problem into parts according to the conditions we need to meet in order to solve the complex problem. In Maimon's example of *Elements* I.1, the two necessary conditions are the sought triangle and

---

166 In the fifth rule of his *Regulae* Descartes writes: "We shall be following this method exactly if we first reduce complicated and obscure propositions step by step to simpler ones" (*CSM*, Vol. 1, p. 379–380). Another example is the second rule of his *Discourse on the Method:* "to divide each of the difficulties I examined into as many parts as possible and as may be required in order to resolve them better" (*CSM*, Vol. 1, p. 18).
167 The use of defining conditions in order to solve a problem appears in rule thirteen of Descartes' *Regulae*. (*CSM*, Vol. 1, p. 435).

equilaterality. The first two simple problems are meant to fulfil the necessary condition of equal sides and the requirement of the number of sides to be three. The third simple problem is meant to ensure that we arrive at a triangle. So even if a complex problem is broken down into its parts, without first intentionally implementing the method of analysis of conditions, the conditions of the problem have to be taken into consideration. However, not all conditions need to be attended to in order to perform the second kind of analysis. In Maimon's own example of *Elements* I.1, there is no need to give an account of whether the "givenness" of the straight line *AB* is a true or pseudo-condition in order to break this problem down in the way Maimon himself proposes.

Maimon's example is productive in that the third problem highlights the need to show that the two radii do in fact meet. Thus, he points to a very important and well known weakness in Euclid's proof of *Elements* I.1. However, there is a problematic element in his application of this method. He solves the first simple problem ("to find two lines which are equal to a given line, and therefore to each other"; *Ueber den Gebrauch*, p. 27) as if it were independent of the original complex problem. In his solution, he finds two lines that are equal to the given straight line (two radii in the same circle) which are connected to the given line at the same endpoint (which is also the center of the circle). This solution solves the first simple problem, but at least one of the two found lines will not take part in the rest of the proof since only one radius in this circle (other than the given line) will serve as a side of the required equilateral triangle. Therefore, the second simple problem does not immediately follow the solution of the first problem. He does not accompany his solutions with diagrams, but were he to present a diagram for the first solution it would be similar to this:

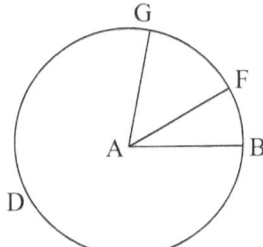

Fig. 4

In the second problem ("to connect the two found lines with the given line on both of its endpoints"; *Ueber den Gebrauch*, p. 27), the reference to "the two found lines" is problematic: the first option is that "the two found lines" means the two found lines in the first solution (*AF, AG*). In this case, the two found lines are radii of the same circle and together these two lines are connected to the given line only at one of its endpoints. Thus, they cannot together meet the condition that the given line is connected at both its endpoints. The second option is that "the two found lines" does not relate to the two radii from the first problem (that are both radii of the first circle), but to one radius from the first circle (*AF*) and one radius from the second circle (*BH*). However, the second circle is constructed only in the solution of the second problem.

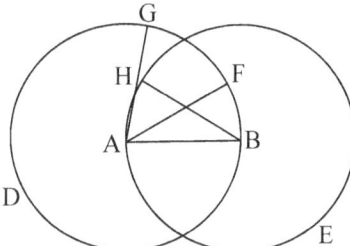

Fig. 5

This problem, arising from Maimon's division of the original complex problem into three simple problems and three solutions, brings our attention to something that is contrary to his initial intention: in some cases, the simple problems are related such that they cannot be reduced to two different parts. Much like the notion of interaction between conditions in science, here too we find interaction of conditions (or interaction between the "simple" problems). I propose altering the first two problems as follows: 1) to find a line that is equal to a given straight line and is attached to it at one of its endpoints; 2) to find another line that is equal to the given straight line and is attached to its other endpoint. In this case, the solution to the first problem is similar to Maimon's and Euclid's solutions: we draw a circle by using the given straight line as its radius and thus can find another radius of this circle so it meets the conditions of both being equal to the given line and being attached to one of its endpoints. The solution to the second problem also follows Maimon's and Euclid's solutions: we draw a second circle by using the given straight line as its radius. Thus, we find another straight line (a radius of the second circle) that meets both requirements of being equal to the given line and being attached to its other endpoint. It is here that the interaction between the two simple problems is found. The first problem can be presented and solved independently of the second. But the second problem can only be posed and solved after the first, as we have to relate to the first problem and its solution in order to know at which endpoint to construct the second circle. In the first solution, we have two possibilities for constructing the circle so that the found line meets the given line at an endpoint. But in the second problem we only have one possibility of determining the center of the second circle at the other endpoint that is not the center of the first circle. Also, the condition "to be connected to one of the given line's endpoints" is derived from a condition that appears only in the complex problem, which is the condition that the form we construct is a triangle. Considering the relations between the simple problems we formulate – whether they depend on each other and if so, what the specific term that creates this dependency is – can help us advance towards finding new solutions and proofs.

As for the third simple problem ("They shall be connected with each other at its other endpoint, so that an equilateral triangle arises"; *Ueber den Gebrauch*, p. 27), Maimon does not specify a solution, but only mentions that the two lines meet where the two circles intersect (*Ueber den Gebrauch*, p. 28).

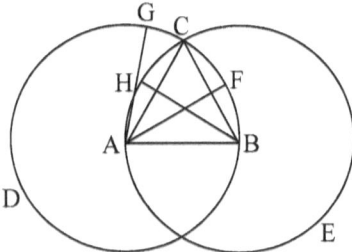

Fig. 6

He does not address the question of whether the two circles meet at one point, nor does he present a full solution to this problem. He does not mention whether he has any knowledge of previous criticisms about whether and how the two radii meet. Nevertheless, it is very plausible that he was familiar with Proclus' commentary of *Elements* I.1 where Proclus discusses Zeno's suggestion that the two constructed lines might have a segment in common and thus be shorter than the given straight line. To this Proclus adds that we need to verify that the two circles also have no common segments (Proclus 1789, p. 27). Maimon's choice of *Elements* I.1 as his example may originated in the wish to highlight that the two lines meet at their endpoints and share no segment. He succeeded in shedding light on this issue, but at the cost of presenting two "simple" problems which are in fact interdependent.

Maimon's solution to *Elements* I.1 stands out from other solutions put forward in contemporary editions of the *Elements*. First, as he himself points out, his way of arriving at a solution reflects the manner in which the inventor arrived (or could have arrived) at the sought solution. The best example of this reconstruction is the first simple problem presented by Maimon. The problem was solved with the construction of only one circle. Although he began by considering two radii of this circle, one of these lines would not be used in the final solution, or even in the solution of the third simple problem. The redundancy involved in finding a line that will not be used later on is typical of the process of searching for solutions in the work of a methodical inventor. It is not typical of a teacher's presentation of a solution to his student. In textbooks, such as Kästner's textbooks used in Germany at that time, the steps that led the inventor to arrive at the solution are not mentioned, only the solution in its final form.[168] The second point of difference between Maimon's solution and other solutions presented in contemporary editions of the *Elements* has to do with the intersection of the circles. In Maimon's solution, first the two straight lines are found equal to a third given line and only then a second circle is constructed. In contemporary editions of the *Elements*, such as those by Simson, Lorenz and Kästner, the proofs begin with the construction of two circles. Although there are differences between their solutions, all three solutions begin with constructing two circles that intersect in the point C. From the point C they draw straight lines (CA, CB) to connect the point of intersection with the given line (AB) (Simson 1811, p. 16; Lorenz 1824, p. 5; Kästner 1786, p. 179–180). Maimon's reordering of the problem and solution had the potential to lead him to ask whether the two circles do in fact meet, but it did not have that effect. This question was raised only in the following century.[169]

In order to examine the connection between the analysis of conditions and the method of simplifying problems and propositions, let us consider proposition 39 of *Data*: "If each of the sides of a triangle be given in magnitude, the triangle is given in form" (Taisbak 2003, p. 119). This proposition corresponds to *El-*

---

[168] Thomas Morel's doctoral dissertation includes a list of courses taught at the University of Leipzig in the last decades of the eighteenth century, together with the textbooks then in use. In the 1780s and 1790s, Kästner's books were frequently used in mathematics courses (Morel 2013, p. 454–469). For instance, in 1780–81 Hindenburg taught a course titled *Arithmetic, geometry and trigonometry*, using Kästner's *Anfangsgründe der Arithmetik, Geometrie, ebenen und sphärischen Trigonometrie, und Perspectiv. Der mathematischen Anfangsgründen ersten Theils erste Abtheilung* (1758) as the accompanying textbook (Morel 2013, p. 454). In 1791, Rüdiger taught a course titled *Elements of Arithmetic and Geometry* using Kästner's *Anfangsgründe der angewandten Mathematik: Zweither Theil, Zweyte Abtheilung: Astronomie, Geographie, Chronologie und Gnomonik* (1781; Morel 2013, p. 460).
[169] For instance, in 1893 Killing raised the question of whether the two circles meet at all (Heath 1956, Vol. I, p. 242).

ements I.22.[170] While Proposition *Dt.* 39 involves a triangle given in magnitude and form, its proof includes more elements than stated in the proposition.[171] It includes the construction of two circles and proving the position of the sides of the triangle and is accompanied by the following diagram:

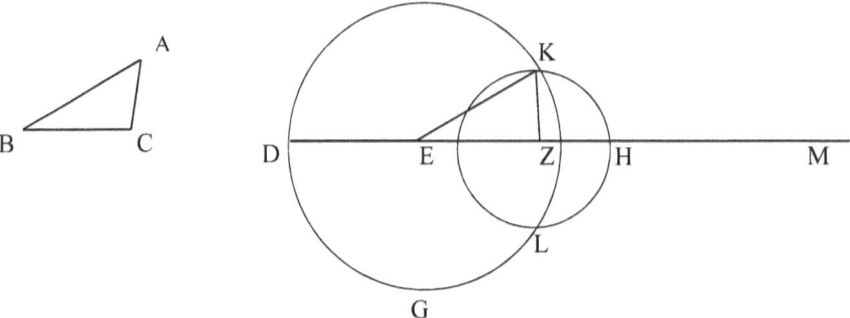

Fig. 7. Based on a diagram from Taisbak 2003, p. 120.

In order to prove that the triangle, which is given in magnitude, is given also in form, we are required to prove that it is given in position. In order to prove it is given in position we need to prove that *KE, EZ, ZK* are given in position. We can break down the complex proposition to the following simple propositions, each one requiring a separate proof. My division of the proposition into simple problems tries to imitate a possible "way of invention" of the proof, so it does not fol-

---

[170] Taisbak compares the proofs of the two propositions, *Dt.* 39 and *Elements* I.22, remarking that they are almost identical (Taisbak 2003, p. 250 – 251).
[171] I am using the proof of *Dt.* 39 as it appears in Taisbak's translation: "For, let each of the sides of triangle *ABC* be given in magnitude; I say that triangle *ABC* is given in form. For, let a straight line given in position *DM* have been set out [Axiom 0*], terminated at *D*, infinite in the other direction, and let *DE* lie equal to *AB*; and *AB* is given [in magnitude]; therefore *DE* is also given [Def. 1], but in position, too [for *DM* is given in position]; and *D* is given, therefore *E* is also given [*Dt* 27]. And [let] *EZ* [lie] equal to *BC*; and *BC* is given, therefore *EZ* is also given, but in position, too; and *E* is given, therefore *Z* is also given. And [let] *ZH* [lie] equal to *AC*; and *AC* is given, therefore *ZH* is also given, but in position too; and *Z* is given, therefore *H* is also given. And with center *E* and radius *ED* let the circle *DKG* have been described; then the [circle] *DKG* is given in position [Def. 6]. Again with center *Z* and radius *ZH* let the circle *HKL* have been described; then the [circle] *HKL* is given in position. And the circle *DGK* is given in position, too; therefore the point *K* is given [*Dt* 25]. And each of the points *E* and *Z* is also given; therefore each of *KE, EZ, ZK* is given in position and in magnitude [*Dt* 26]; therefore triangle *KEZ* is given in form [Def. 3*A]. And it is equal and similar to triangle *ABC* [I.8, I.4, VI. Def. 1]; therefore triangle *ABC* is given in form [Def. 3*B]" (Taisbak 2003, p. 119 – 120).

low the exact order of Euclid's proof. Instead, it is divided in order to better represent whether there is dependence between simple problems or conditions:

(1) triangle ABC is given in form if there is another triangle similar to it given in form.

This proposition can be proven by using Def. 3*B' ("A triangle is given in form if it is similar to one that is given in form"; Taisbak 2003, p. 33). We will construct a similar triangle (triangle KEZ).

(2) Triangle KEZ is given in form if it is given in magnitude and position.

This proposition can be proven by using Def. 3*A ("A triangle is given in form if its vertices are given in position"; Taisbak 2003, p. 33) and showing that the sides of triangle KEZ (KE, EZ, ZK) are given in position and in magnitude.

(3) The sides of triangle KEZ (KE, EZ, ZK) are given in position and magnitude, if their extremities K, E, Z are given in position.

This proposition can be proven by using Dt. 26 ("If the extremities of a straight line be given in position, the line is given in position and in magnitude"; Taisbak 2003, p. 99).

(4) E, Z, K, are given in position.

To prove (4) we need to prove the following propositions, using this construction: Let the straight line DM be given in position, terminated at D and infinite in the other direction (the construction is possible due to Axiom 0*: "Any point or magnitude in the plane may be (taken and) appointed given"; Taisbak 2003, p. 25).

(4.1) E is given in position:

The proof of this proposition demands that we first prove:

(4.1.1) DE is given in magnitude and in position.

This proposition can be proven by constructing DE equal to AB. AB is given in magnitude and DM is given in position, therefore DE is given in magnitude and in position (in order to prove it is given in magnitude we use Def. 1: "Given in magnitude is said of figures and lines and angles for which we can provide equals"; Taisbak 2003, p. 17).

Since DE is given in magnitude and in position and since D is given in position, E is also given in position (using Dt. 27:"If one extremity of a straight line given in position and in magnitude be given, the other will also be given; Taisbak 2003, p. 100).

(4.2) Z is given in position.

The proof of this proposition demands that we first prove:

(4.2.1) EZ is given in magnitude and in position.

This proposition can be proven by constructing EZ equal to BC. BC is given in magnitude, therefore, EZ is given in magnitude and in position.

Since *EZ* is given in magnitude and in position, and since *E* is given in position, *Z* is also given in position

(4.3) *K* is given in position.

The proof of this proposition demands we first prove:

(4.3.1) *ZH* is given in magnitude and in position.

This proposition can be proven by constructing *ZH* equal to *AC*. *AC* is given in magnitude therefore *ZH* is given in magnitude and in position.

Since *ZH* is given in magnitude and in position and since *Z* is given in position, *H* is also given in position.

Now it is left to be proven that:

(4.3.2) Circle *DGK* is given in position.

This proposition can be proven by constructing a circle *DKG* with the center *E* and radius *ED* (using Def. 6: "And a circle is said to be given in position and in magnitude if its centre is given in position and its radius in magnitude"; Taisbak 2003, p. 34).

(4.3.3) Circle *HKL* is given in position.

This proposition can be proven by constructing a circle *HKL* with the center *Z* and radius *ZH*.

Since circles *DGK* and *HKL* are given in position, then point *K* is given in position (according to *Dt.* 25: "If two lines given in position cut one another, their point of section is given in position"; Taisbak 2003, p. 93).

Breaking down the proposition (and proof) into simpler components is done by identifying the various conditions of the problem and of the solution. Being given in position is a necessary condition of the proof, but one that is not indicated as a condition in proposition *Dt.* 39. Identifying this true condition of the proof (proving that triangle *KEZ* is given in position), is necessary for the second kind of analysis. The simple proposition: "to prove that *KE* is given in position" is not derived directly from the initial proposition ("If each of the sides of a triangle be given in magnitude, the triangle is given in form"); but we arrive at it only by using the first kind of analysis. Reconstructing the problem and the solution in this manner emphasizes the dependency of some conditions (or "simple problems") on others. For instance, that "triangle *ABC* is given in form" depends on whether *E*, *Z*, *K* are given in position, or that "*E* is given in position" depends on whether triangle *ABC* is given in magnitude. Also, in (4.3), we come to realize that the question of whether the two lines cut each other is relevant to proposition *Dt.* 25 ("If two lines given in position cut one another, their point of section is given in position"; Taisbak 2003, p. 93) and therefore also to proposition *Dt.* 39.

Applying the analysis of conditions to this problem shows that the condition "*DM* is infinite on the side of *M*" is a pseudo-condition. For the reason that what is required is that *DM* would be greater than *DE* and twice *EZ* taken together (thus ensuring the ability to construct both circles *DGK* and *KHL* cutting *DM*), not that *DM* is infinite on the side of *M*. Since the sides are given in magnitude, there is no reason that we would not be able to fulfill such a condition. This is especially so if we consider that in the corresponding proposition, *Elements* I.22, we have a necessary condition that any two straight lines taken as sides of the triangle would be together greater than the third straight line serving as a side.[172] Just as we can ensure that this condition holds, there is no reason that we could not add the condition that the straight line *DM* (given in position and hence in magnitude) would be greater than *DE* and twice *EZ* taken together.[173] Infinity does not add to the solution other than by ensuring the ability to construct the required circles and we can meet this requirement with finite measures as well.

4.1.3 Analysis of the Cases of a Problem or a Proposition

A problem or a proposition can have several cases. Therefore, claims Maimon, we need to solve or prove all these particular cases in order for the problem or proposition to be general [*allgemein*]. He offers the following explanation for this need: mathematical propositions are not analytic, since they cannot be proven using concepts alone.[174] Were they analytic, we could have proven them universally since what applies to the universal [*Allgemeinen*] must also apply to all the particular cases subsumed under it.[175] Mathematical propositions

---

[172] "Out of three straight lines, which are equal to three given straight lines, to construct a triangle: thus it is necessary that two of the straight lines taken together in any manner should be greater than the remaining one" (Heath 1956, Vol. I, p. 292). Although this condition does not appear in *Data*, prop. 39, it is a necessary condition for the original triangle (*ABC*), given in magnitude.

[173] Taisbak does not mention *Dt.* 26 ("If the extremities of a straight line be given in position, the line is given in position and in magnitude"; Taisbak 2003, p. 99) in relation to the given line *DM*, but it is applicable here: since both *D* and *M* are given in position (they lie on the straight line DM given in position), *DM* is given in magnitude.

[174] Here Maimon refers to analysis in its narrow sense: analysis of the concept, grounded in the principle of contradiction alone.

[175] Maimon uses the term *allgemein* to describe both general and universal propositions. I chose to translate *allgemein* as *universal* in a manner similar to Aristotle's use of the term: universal is a form, independent of its particular instances, apprehended by the intellect. *General*, on the other hand, was translated in a manner similar to Locke's use of the term: a general proposition is the outcome of abstraction. Thus, the more general proposition in the analytic process is referred here as *universal* while the general proposition in the synthetic process is referred to as *general*.

are synthetic since their proofs are not derived from concepts but from construction. In synthetic knowledge, all the particular cases need to be proven in order for the general proposition to arise. Therefore, he concludes, there is a need to prove all the particular cases with their own particular constructions in order for the propositions to be proven generally (*Ueber den Gebrauch*, p. 28 – 29; *Das Genie*, p. 379). Maimon's third kind of analysis is generalization since we begin by proving the particular propositions and arrive at the more general proposition. The direction of this method is opposite to that of Leibniz's method of universalization, where he finds a single operation that can then be applied to several different cases (Leibniz 1961a, p. 97). He finds and proves the universal rule, thereby proving all the particular cases subsumed under it. Even though Maimon regards generalization as a synthetic process and universalization as analytic, the method of finding all the possible cases of a problem or a proposition is considered by him as analysis. For the reason that these cases are found by decomposing a given proposition into its conditions and cases, i.e. by analysis.

To assert the generality of a proposition (or a problem), we need to consider all its cases. Maimon writes that "A problem or a proposition can have several cases, each of which needs to be resolved in a particular way, or must be proven, if the problem or the proposition should be *general*" (*Ueber den Gebrauch*, p. 28). This is a source for another method of invention, that was not mentioned by Maimon: analysis of the number of cases of a problem or a proposition. We should not only show the cases of a given problem, but also that there are no other cases that are yet unproved. An example of such a method is Euclid's proposition at the end of Book XIII of *Elements:* "no other figure, besides the said five figures, can be constructed which is contained by equilateral and equiangular figures equal to one another" (Heath 1956, Vol. III, p. 507).[176] This proposition and its proof can serve as an analysis of the number of cases of the general problem: "to construct a figure that is contained by equilateral and equiangular figures equal to one another".

Maimon presents *Elements* III.20 as an example: "the angle at the center of a circle is twice the size of the angle standing on the periphery on the same arch". In order for the proposition to be general, we need to prove all three particular cases, each requiring its own construction (*Ueber den Gebrauch*, p. 29). Maimon mentions in passing that Euclid's proof of *Elements* III.20 gives an account of several cases, but he elaborates neither on the details of this proposition nor

---

[176] This proposition appears at the end of Book XIII, after Proposition 18. The five mentioned figures are pyramid, octahedron, cube, icosahedron and dodecahedron (proven in Prop. 13 – 17).

on the proof of these particular cases.[177] In the proof of *Elements* III.20[178], Euclid demonstrates that angle *BEC* is double the angle *BAC* by producing line *AE* to *F*, showing that angle *BEF* is double the angle *EAB*, and that the angle *FEC* is double the angle *EAC*. In order to solve the second case, he constructs the line *BD* and produces the line *DE* to *G*. It is proven that the angle *GEC* is double the angle *EDC* by demonstrating that angle *GEB* is double the angle *EDB* and that angle *BEC* is double the angle *BDC* (Heath 1956, Vol. II, p. 46–47).

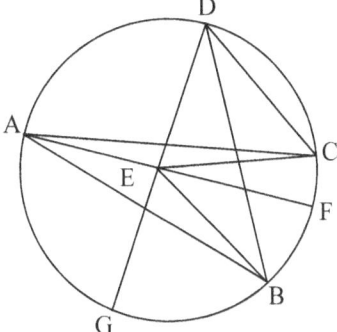

Fig. 8. Based on a diagram from Heath 1956, Vol. II, p. 46.

In the first case, the angle at the center of the circle is less than a right angle (angle *BEC*). In the second case, the angle at the center of the circle is greater than a right angle and less than two right angles (angle *GEC*). In his commentary, Heath presents Heron's extension of this proposition to the case where the angle at the center is greater than two right angles (Heath 1956, Vol. II, p. 47).[179] Heath

---

177 In *Giv'at Hamore*, Maimon mentions proposition *Elements* III.20 together with its first case of the problem and proof. He does not, however, elaborate on other cases or on the subject of cases of a problem in general. He uses this proposition to show how even if a premise is impossible or false and consequently the conclusion is impossible or false, the inference itself remains possible and true. For instance, were we to assume the false premise "in a triangle, the exterior angle is equal to the two interior and opposite angles taken together with half of their sum" as true, consequently we would arrive at the false conclusion: "in a circle, the angle at the center is three times of the angle at the circumference (when the angles have the same circumference as base)". However, the inference itself, i.e. the action of connecting truths according to the rules of the understanding, is possible and true (*Giv'at Hamore*, p. 135–6).
178 *Elements* III.20: "In a circle the angle at the center is double of the angle at the circumference, when the angles have the same circumference as base" (Heath 1956, Vol. II, p. 46).
179 "The angle which is at the center of any circle is double the angle which is at the circumference of it *when one arc is the base of both angles*; and *the remaining angles which are at the*

states that since an angle in the Euclidean sense is that which is less than two right angles, his proof consists of the sum of certain "angles" into an "angle greater than two right angles" (Heath 1956, Vol. II, p. 47).[180] Heron's proof demonstrates that "any angle in the segment BAC is half of the angle BDC; and the sum of the angles BDG, GDF, FDC is double any angle in the segment BEC" (Heath 1956, Vol. II, p. 47–48).

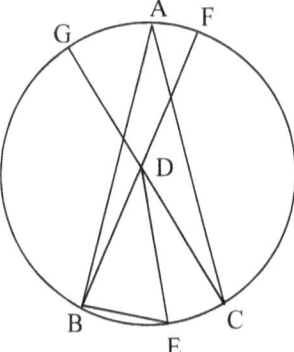

Fig. 9. Based on a diagram from Heath 1956, Vol. II, p. 48.

Heron's additional case is a result of his analysis of the condition that an angle is less than two right angles, a condition arising from the Euclidean definition of an angle. This could be a source for another fruitful method: the method of amendment of a definition and consequently the presentation of a new case of a problem or proposition (and its solution or proof). In this case, Heron's amendment was to add angles (each less than two right angles, complying with Euclid's definition) so that together they are greater than two right angles. It is an example of what is entailed in "taking small steps", as suggested by Maimon, steps the methodical inventor can account for (*Das Genie*, p. 373).

To closely examine this method, I would like to present Euclid's proof of *Elements* I.35, following the additional cases added by Proclus. Euclid presents only one case in his proof of proposition I.35, "Parallelograms which are on the same base and in the same parallels are equal to one another" (Heath 1956, Vol. I, p. 326). Parallelograms *ABCD* and *EBCF* are given on the same base *BC* and in the same parallels *AF* and *BC*:

---

center, and fill up the four right angles, are double the angle at the circumference of the arc which is subtended by the [original] angle which is at the center" (Heath 1956, Vol. II, p. 47).
180 Def. 8 of *Elements* is "A *plane angle* is the inclination to one another of two lines in a plane which meet one another and do not lie in a straight line" (Heath 1956, Vol. I, p. 153).

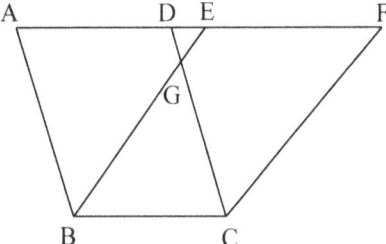

Fig. 10. Based on a diagram from Heath 1956, Vol. I, p. 327.

Euclid's proof of the equality of the parallelograms is based on showing that the two triangles *ABE* and *DCF* are equal. By subtracting triangle *DGE* from both triangles, he shows that trapezium *ABGD* is equal to trapezium *EGCF*. Finally, by adding triangle *GBC* to each trapezium it is shown that parallelogram *ABCD* is equal to parallelogram *EBCF* (Heath 1956, Vol. I, p. 326–327). In his commentary, Proclus mentions that Euclid has chosen a difficult case and adds two more cases of his own to this proposition. In the first case, the parallelograms *ABCD* and *BDCE* share the base *BD* and meet at point *C* (Proclus 1789, p. 181).[181]

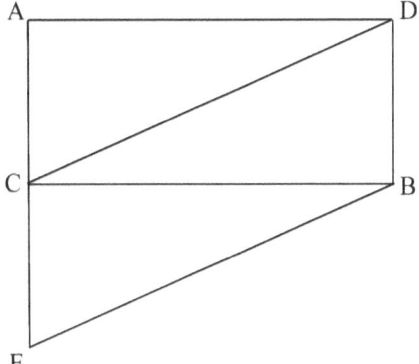

Fig. 11. Based on a diagram from Proclus 1789, p. 182.

---

181 Proclus proves this case as follows: parallelograms *ABCD* and *BDCE* share the base *BD*. Let the side *CD* be the diameter of parallelogram *ABCD* and the side *CB* the diameter of parallelogram *BDCE*. Since the diameter bisects the area of the parallelogram [*Elements*, I.34], the area of triangle *BCD* is half the area of parallelogram *ABCD* and it is half the area of parallelogram *BDCE*. Therefore, the areas of parallelograms *ABCD* and *BDCE* are equal (Proclus 1789, p. 181–182).

In the second case added by Proclus, the parallelograms *ADBE* and *DBFC* are given. They share the base *BD* and the side *AE* is cut by the side *DC* (Proclus 1789, p. 182).[182]

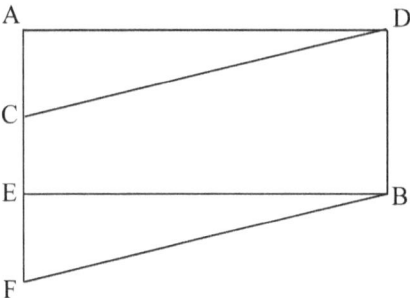

Fig. 12. Based on a diagram from Proclus 1789, p. 182.

After the presentation of the last case, Proclus writes: "And here you may observe that these are the only three cases. For the side *DC*, either cuts the side *EB*, according to the position of the elementary institutor; or it falls on the point *C*, as in the penultimate description: or it cuts the line *AE*, according to the present supposition. And thus the theorem is shewn to be true according to all its cases" (Proclus 1789, p. 182–183). This echoes Maimon's description of the method and could have served as a source of influence. Proclus' proof of *Elements* I.35 is a good example of both analysis of cases of a proposition and analysis of conditions. Examining the conditions of the proposition, we can determine that the true conditions are that the two forms are parallelograms and that they share one side. Further examination can lead us to another condition, based on the type of relation between the two sides of the two parallelograms opposite to the shared side: they either share a segment, do not share a segment, or meet at one point. This analysis of the conditions of the problem is helpful in the analysis of the cases of the problem since the different conditions can highlight what the possible cases are. One may ask: how can we know be-

---

[182] Proclus' proves this case as follows: the parallelograms *ADBE* and *DBFC* are given and the side *AE* is cut by the side *DC*. Both parallelograms share the base *BD*, therefore the side *AE* is equal to the side *CF*. We then subtract the common line *CE* from both sides, so that *AC* is equal to *EF*. Since *AD* is parallel to *EB*, the side *AD* is equal to the side *EB* and angle *CAD* is equal to angle *FEB*. The base *CD* is equal to the base *FB*, and triangle *ADC* is equal to triangle *EBF*. Finally, the common trapezium *CDBE* is added. Proclus concludes that the whole *ADBE* is not unequal to the whole *DBFC* (Proclus 1789, p. 182). These three cases are also presented in Simson's version of the *Elements* (Simson 1811, p. 40).

forehand whether it is required to prove the particular cases or whether the presented case can be taken as general? Maimon does not refer to this question. Based on Maimon's account of problems as composed of simple problems, themselves considered to be conditions (as presented in the second kind of analysis), we can assume that the criterion Maimon would have given for presenting several cases of a problem would be whether there is a possible change of condition.[183]

Changing one condition of a problem (or proposition) is what distinguishes the analysis of the cases of a problem (third kind of analysis) from the analysis of the cases of a solution (fifth kind of analysis). In analysis of the cases of a problem we present a new solution involving both a new condition and a new construction. In Euclid's proof of *Elements* III.20, for instance, the new condition (together with its new construction) is that the angle at the center of the circle is less than a right angle in the first case. In the second case, it is that the angle at the center of the circle is greater than a right angle, but less than two right angles. In *Elements* I.35, the new cases of the problem are a result of a change in the condition according to the relation between the two sides of the two parallelograms opposite to the shared base (whether they meet, share a segment or neither share a segment nor meet). Per contra, analysis of the cases of the solution (fifth kind of analysis) requires only a new construction in the solution and does not change any conditions in the given problem (*Ueber den Gebrauch*, p. 31).

4.1.4 Analysis of the Object

Describing his fourth kind of analysis, Maimon states that sometimes the sought premises of a solution or proof do not refer immediately to the given object of a problem or proposition. In these cases, the given object of a problem or proposition, to which the premises refer immediately, is different than the sought object to which the conclusion of the proof or the solution refers. In order to solve the problem or prove the proposition, the given object needs to be transformed into the sought object by using alterations such as increase or diminution. Without this analysis of the object, Maimon writes, the invention of a solution or a proof is impossible (*Ueber den Gebrauch*, p. 29–30; *Das Genie*, p. 379–

---

[183] A similar answer is given by Manders: After presenting a problem for which Apollonius presented 87 cases, he mentions that case distinctions are not made in Euclidean geometry when a particular case can be substituted in the proof by another particular case. The distinctions are made only when there are preconditions of possibility that demand a different diagram type (Manders 2008, p. 104–105).

380). In order to prove *Elements* I.5 ("In every isosceles triangle the angles at the base are equal to each other"), Maimon poses the question: how can it be shown that these angles are equal to each other, based on previously proven propositions? This question leads to *Elements* I.4 where equality of angles is found.[184] However, in *Elements* I.4 the proposition and proof refer to equal angles that are in two different triangles whereas in *Elements* I.5 the angles are in the same triangle. Therefore, in order to be able to apply *Elements* I.4 to the proof of *Elements* I.5, the given object needs to be transformed so that the objects of propositions I.4 and I.5 can be connected. The equality of the angles at the base of the given isosceles triangle of *Elements* I.5 is proven by showing that these angles are also angles in two different triangles, consequently allowing the application of proposition I.4 to demonstrate their equality (*Ueber den Gebrauch*, p. 30 – 31). However, a closer look at the details of *Elements* I.5 raises a few difficulties in Maimon's description of the method and its example. Primarily, the proposition includes more than the angles at the base of the isosceles triangle, it includes also the angles under it: "In isosceles triangle the angles at the base are equal to one another and, if the equal straight lines be produced further, the angles under the base will be equal to one another" (Heath 1956, Vol. I, p. 251). The object of an isosceles triangle *ABC* is given, the side *AB* is equal to side *AC*. The proof begins by producing two straight lines: *BD* is produced further in a straight line with *AB*, and *CE* is produced further in a straight line with *AC* (Heath 1956, Vol. I, p. 251).[185]

---

184 *Elements* I.4: "If two triangles have the two sides equal to two sides respectively, and have the angles contained by the equal straight lines equal, they will also have the base equal to the base, the triangle will be equal to the triangle, and the remaining angles will be equal to the remaining angles respectively, namely those which the equal sides subtend" (Heath 1956, Vol. I, p. 247).
185 The proof continues as follows: Point *F* is taken at random on *BD*, and *AG* is cut off from *AE* equal to *AF*. Then the straight lines *FC* and *GB* are joined. Since *AF* is equal to *AG*, and *AB* is equal to *AC*, and since the angle *FAG* is a common angle, it is shown that triangle *AFC* is equal to the triangle *AGB*, by applying *Elements* I.4. Therefore, the base *FC* is equal to the base *GB*, angle *ACF* is equal to the angle *ABG*, and angle *AFC* is equal to angle *AGB*. Since the whole of *AF* is equal to the whole of *AG*, and *AB* is equal to *AC*, the remainder *BF* is equal to the remainder *CG*. Euclid then shows that triangle *BFC* is equal to triangle *CGB*, since *BF* is equal to *CG* and *FC* is equal to *GB*, and *BC* is a common base. Therefore, the angle *FBC* is equal to the angle *GCB* (thus proving the equality of the angles under the base), and the angle *BCF* is equal to the angle *CBG*. Since the whole angle *ABG* is equal to the angle *ACF*, and since angle *CBG* is equal to angle *BCF*, the angle *ABC* is equal to angle *ACB* (thus proving the equality of the angles at the base; Heath 1956, Vol. I, p. 251–252).

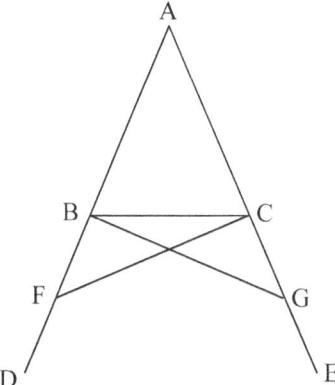

Fig. 13. Based on a diagram from Heath 1956, Vol. I, p. 251.

The attempt to describe the given object of this altered proposition raises the following question: is the given object in *Elements* I.5 the isosceles triangle *ABC* alone and the production of *BD* and *CE* a part of the analysis of the object, or is the object *DBACE* the original given object to be altered and analyzed? Let us use Proclus' division of the parts of a proof for this proposition[186]: the *exposition* (*ekthesis*) is "Let *ABC* be an isosceles triangle having the side *AB* equal to the side *AC*; and let the straight lines *BD*, *CE* be produced further in a straight line with *AB*, *AC*" and the determination (*diorismos*) is "I say that the angle *ABC* is equal to the angle *ACB*, and the angle *CBD* to the angle *BCE*" (Heath 1956, Vol. I, p. 251). The construction (*kataskeuê*) is "Let a point *F* be taken at random on *BD*; from *AE* the greater let *AG* be cut off equal to *AF* the less; and let the straight lines *FC*, *GB* be joined" (Heath 1956, Vol. I, p. 251). Applying Proclus' division to this proposition, it seems that the original given object is *DBACE* and the amendment of the given object includes only the production of *FC* and *BG*. This is in contrary to what Maimon regards as the given object – triangle *ABC*. In other words, the line between what is given in the proposition and what belongs

---

[186] Proclus divides proofs of the *Elements* into six parts. The propositions are first presented as general statements (called *enunciation* or *protasis*). These are followed by a given particular instance (*exposition* or *ekthesis*) and what is sought in the particular instance (*determination* or *diorismos*). Following this is the construction that is conducted on the given in order to arrive at what is sought (*construction* or *kataskeuê*), the proof (*demonstration* or *apodeixis*) and the conclusion in terms of the particular instances (*conclusion* or *sumperasma*) (Harari 2004, p. 109; Proclus 1789, p. 20). As Mueller indicates, from a modern standpoint, everything following the *protasis* saving the *diorismos* is proof, whereas in Euclidean terms only the *apodeixis* means 'proof' (Mueller 1981, p. 11).

to the method of analysis of the object is not always clear. Perhaps the reason for the difference in what is considered as the given object depends on the fact that Maimon only quotes the first part of the proposition.[187] This choice is puzzling since first the equality of the angles under the base is proven and only then is the equality of the angles at the base proven. There are three possible reasons for not mentioning the second part of the proposition. First, Maimon could have wanted to make his argument as short as possible and thus presented only the first part of the proposition. Second, in Proclus' commentary to this proposition, he mentions that this part of the theorem has no use in the *Elements* (Proclus 1789, p. 51), and it is possible that this, too, affected Maimon's decision not to include it. Lastly, it might be the result of Maimon's application of the second kind of analysis – the method of simplifying propositions and problems – to this example. Maimon mentions that in order to conduct analysis of the object, one should first conduct the first two kinds of analysis (*Ueber den Gebrauch*, p. 32). In this he might have been influenced by Proclus' commentary on *Elements* I.5, which begins with the statement that some theorems are composite and some are simple. Proclus mentions that *Elements* I.5 is composite (Proclus 1789, p. 50) and, as an example of simple propositions, he presents "Every isosceles triangle has the angles at the base equal" (Proclus 1789, p. 49). In this case, Proclus and Maimon share the opinion that the first part of the proposition can be regarded as simple. However, unlike Maimon who mentions only the division of complex propositions into simple ones, Proclus also believes that in some cases it is better for analysis that the proposition be considered as composite and not simple.[188] It is also plausible that Maimon was influenced by Proclus, who raises the question of the need to prove the equality of angles under the base and presents two proofs of only the first part of the proposition (Proclus 1789, p. 51–52).

Although *Data* is traditionally referred to as a book "in the treasury of analysis", it includes the use of construction in some of its proofs and more specifically, the method of analysis of the object. For instance, *Dt.* 31: "If from a given point a straight line given in magnitude be drawn to meet a straight line given in position, the line drawn is also given in position" (Taisbak 2003, p. 104).

---

187 Both Simson's English translation (1756) and Lorenz's German translation (1781) include the complete version of the proposition.

188 Proclus writes: "In general geometers have made such composite propositions both with a view to brevity and for the purpose of analysis; for often things left uncompounded do not lend themselves to analysis and only when put together provide an easy way of getting back to first principles" (Proclus 1992, p. 246: see also p. 192).

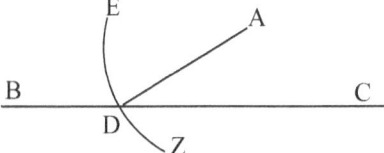

Fig. 14. Based on a diagram from Taisbak 2003, p. 104.

The straight line *DA* is given in magnitude and is drawn from the given point *A* to the straight line *BC*, which is given in position. We are required to demonstrate that the line *DA* is given in position as well. Proving the sought cannot be done without the use of previous propositions and definitions and their application requires a change in the given object. Hence, circle *EDZ* is constructed.[189]

The principal act of this method is the transformation of the object given in intuition into another object that refers to the proof or solution, using construction. Although based on construction, Maimon intends for this method to be used in logical analysis. For Maimon, the importance of this method lies not in the object that is analyzed but in the process of analysis itself.[190] He views construction as secondary to inference (i.e. the faculty of imagination as secondary to the faculty of reason). Still, the method itself is the action of the faculty of imagination in intuition and not inference. It is different from the analysis of the object described in Maimon's *Logic* (e.g. *Logik*, p. 29) which is not concerned with changing the object but with what new properties we can educe from the given object. In all cases of analysis of the object, the given objects are given in actuality and are not pure symbolic objects.

---

189 In order to show that the line *DA* is given in position, proposition *Dt.* 26 can be used ("If the extremities of a straight line be given in position, the line is given in position and in magnitude"; Taisbak 2003, p. 99). Since point *A* is given, in order to apply this proposition to line *DA*, we are required to prove that point *D* is given in position using *Dt.* 25 ("If two lines given in position cut one another, their point of section is given in position."; Taisbak 2003, p. 93). We cannot use the given straight lines in order to apply this proposition to the given object *ABCD* since it assumes our conclusion that line *DA* is given in position. Hence, the construction of another geometrical magnitude is needed: circle *EDZ* (with center *A* and radius *DA*). Since the center of the circle, point *A*, is given in position and the radius *DA* is given in magnitude, we can apply Def. 6 ("And a circle is said to be given in position and in magnitude if its centre is given in position and its radius in magnitude"; Taisbak 2003, p. 34). Once circle *EDZ* is given in position, we apply *Dt.* 26 to the new object and arrive at the proposition "point *D* is given in position" and consequently at the sought proposition "*DA* is given in position" (Taisbak 2003, p. 104).

190 For instance, he writes: "For *invention*, not the *analyzed object* but the analysis itself should be presented, how it happens step by step, or the *methods* [should be presented], [to show] how this *analysis* should be employed" (*Ueber den Gebrauch*, p. 32).

It is unclear whether Maimon is claiming that this method is indispensable to every invention or whether he refers only to the cases in which the sought conclusion refers to an object different from the one given (*Ueber den Gebrauch*, p. 30).[191] In either case, in Euclid's *Elements* there are demonstrations that do not include further construction beyond constructing the given object and do not include any change in the object, such as in *Elements* I.4. Hence, it can be determined that this kind of analysis is not applied to every proof or solution.

This method was in use before Maimon's time, only it did not carry this name. Maimon's contribution in this respect is not the invention of the method (something for which he did not claim credit) but prescribing the rule of invention: "transform the given object into the sought object, by using alterations such as increase or diminution so that you can apply a given proposition to it". It allows the methodical inventor to consciously look for possible alterations of the given object so that a given proposition can be applied to it, rather than follow such a method unconsciously.

4.1.5 Analysis of the Cases of the Solution

This method is applied to problems alone, not to theorems. In describing this method, Maimon mentions that in some cases we can arrive at the sought solution in more than one way (*Ueber den Gebrauch*, p. 31). However, this declaration and the example he gives are at odds: his description of the method might suggest that it is meant to help us present new solutions, but his example indicates that this method offers only new constructions, following the same method of solution. Maimon's statement, that a solution is complete only once all the ways to solve the problem are presented, is accompanied by the example of *Elements* I.1, claiming that we can construct an equilateral triangle not only above the given line, but also under it (*Ueber den Gebrauch*, p. 31; *Das Genie*, p. 380). This method is the least important one since it does not present a new way of solving the problem, only a duplication of a given solution using a slightly different construction. It is an analysis of the ways all the given possibilities of construction of the sought object can be presented under given conditions, both the conditions of the problem and of the solution.

---

191 On the one hand, Maimon writes: "This I call: *analysis of the object*, without which the *invention* of a solution or a proof is impossible" (*Ueber den Gebrauch*, p. 30; *Das Genie*, p. 380). On the other hand, in *The Genius and the Methodical Inventor*, just before writing the same sentence, he specifically mentions "in this case", referring to the case in which the sought conclusion does not refer to the given object (*Das Genie*, p. 380).

This method is better understood in light of other methods. The method of analysis of the cases of a problem or proposition (third kind of analysis) is based on changing one condition of a problem or proposition so that each case demands its own solution and construction. In the method of analysis of the cases of the solution (fifth kind of analysis), by way of contrast, there is no change in any condition. It only adds more constructions which conform to the same kind of solution. Moreover, Maimon mentions that all the particular cases of a problem or proposition must be solved for the solution or proof to be general [*allgemein*] (*Ueber den Gebrauch*, p. 28), whereas a solution should be presented in all possible ways for it to be complete [*vollständig*] (*Ueber den Gebrauch*, p. 31). Even though not mentioned by Maimon, this difference is based on whether there is a change in the conditions between cases: in the third kind of analysis, each change of condition requires a different solution so that it is necessary to solve or prove all the particular cases for the problem or proposition to be general. In the fifth kind of analysis, however, the solution is more complete in the sense that we represent all possible constructions that fulfill the conditions of the problem and solution, but it is no more general. The universality of the solution is established already by the first case of the solution of the problem. In *Elements* I.1, for instance, by constructing the equilateral triangle above the line we have solved the given problem. This solution is universal in regards to any given straight line regardless of its size and position. Likewise, the solution is indifferent to the direction of the construction of the triangle – whether it is above or below the given line – since all the conditions of these two cases are the same.[192]

---

[192] The universality of a solution is a known characteristic of Greek geometry. Proclus describes how the solution of a particular instance given in magnitude serves as a universal solution of the problem: "Furthermore, mathematicians are accustomed to draw what is in a way a double conclusion. For when they have shown something to be true of the given figure, they infer that it is true in general, going from the particular to the universal conclusion. Because they do not make use of the particular qualities of the subjects but draw the angle or the straight line in order to place what is given before our eyes, they consider that what they infer about the given angle or straight line can be identically asserted for every similar case. They pass therefore to the universal conclusion in order that we may not suppose that the result is confined to the particular instance. This procedure is justified, since for the demonstration they use the objects set out in the diagram not as these particular figures, but as figures resembling others of the same sort. It is not as having such-and-such a size that the angle before me is bisected, but as being rectilinear and nothing more. Its particular size is a character of the given angle, but its having rectilinear sides is a common feature of all rectilinear angles. Suppose the given angle is a right angle. If I used its rightness for my demonstration, I should not be able to infer anything about the whole class of rectilinear angles; but if I make no use of its rightness

Maimon does not define what "case" means, but we can assume it is similar to Proclus' notion: "A 'case' announces that there are different ways of making the construction, by changing the position of the points, lines, planes, or solids involved. Variations in case are generally made evident by changes in the diagram, whereof it is called 'case', because it is a transposition in the construction" (Proclus 1992, p. 212). From Maimon's use of the cases of the problem or proposition and cases of the solution, it can be inferred that Maimon's notion of a case is "any alteration in construction". Thus, presentation of a new construction either under the same given conditions or under changed conditions are both considered as cases.

The third kind of analysis refers to cases of both problems and propositions, whereas the fifth kind refers to cases of solutions alone, not to proofs. The cause of this difference lies in the question of whether the sought is a new object or a new property. According to Proclus, in problems we create new objects and in theorems we find new attributes of a given object (Proclus 1992, p. 201). Using Maimon's terminology, in problems we invent and in theorems we discover. In the third kind of analysis, the different construction of each case is a consequence of a change of condition, which can be performed on propositions as well as on problems. Once this change of condition is conducted, a new object is presented as given. However, once an object is given, we cannot change its conditions. The fifth kind of analysis is applied only to problems since, given an object, we can invent additional objects under the same conditions by using construction, but we cannot discover new properties that were not already found without changing the conditions and thus changing the given object.

Common to the fifth kind of analysis and analysis of the conditions of the possibility of the solution (which is a part of the first kind of analysis, see Section 4.1.1.2) is that both refer to the possibility of constructing objects. They differ in that the first aims at showing all the possible constructions of the sought object that can be presented in actuality, under the given conditions presented in the problem. The latter enables the construction of the sought object as such. It refers to solutions alone and not to proofs since it deals with producing a new ob-

---

and consider only its rectilinear character, the proposition will apply equally to all angles with rectilinear sides" (Proclus 1992, p. 207). Proclus makes a similar claim specifically in regard to the solution of *Elements* I.1: "the three lines therefore are equal, and an equilateral triangle [*ABC*] has been constructed on this given straight line." This is the first conclusion ensuing the exposition, followed by the general conclusion: "An equilateral triangle has therefore been constructed upon the given straight line. For even if you make the line double that set forth in the exposition, or triple, or of any other length greater or less than it, the same construction and proof would fit it" (Proclus 1992, p. 164).

ject. The requirement of verifying that the object is possible is not relevant to theorems since the object is already given as actual, its possibility is derived from its actuality.

In Maimon's analysis of the various ways according to which a problem can be solved, or a proposition proven (the sixth kind of analysis), there is an indefinite number of solutions to a given problem. In the fifth kind of analysis, there is a limited number of cases of a solution. When discussing the method of analysis of the cases of the problem, I have suggested a method of analysis of the number of cases of a problem (Section 4.1.3). Similarly, I suggest a method of analysis of the number of possible cases of solutions. In the case of *Elements* I.1, Maimon mentions that the two cases are the only cases, based on the assumption that the two radii meet at the two points of intersection of the circles. By examining the conditions of the solution, we can determine the number of possible cases of a solution.[193]

Maimon's example of an application of this method, *Elements* I.1, allows us to discover an unmentioned requirement of the method: the ability to construct the different cases simultaneously. This requirement also provides an explanation as to why this method is meant to be applied only to problems and not to theorems. Maimon writes that in order to have the complete solution of the problem presented in *Elements* I.1, we should construct another equilateral triangle under the given straight line: "Thus for example, the 1st problem in Book I receives, by the use of precisely the same solution, two constructions at the same time, in which the equilateral triangle can be constructed both above and under the given line" (*Ueber den Gebrauch*, p. 31). Following Maimon's suggestion, let us take the radii *AF* and *BF* to construct the equilateral triangle *AFB*, to be presented alongside Euclid's given solution, that is, triangle *ABC*.

---

193 Such analysis of the number of possible cases of a solution is found in algebra, embodied in the formulas that indicate how many real roots a quadratic equation has: if the discriminant is positive, the equation has two real roots; if it is zero, there is one real root; and if it is negative, there are no real roots. A similar method is introduced by Descartes in *Geometry* (one of the appendixes of *Discourse on the Method*): "[...] We can determine also the number of true and false roots that any equation can have, as follows: An equation can have as many true roots as it contains changes of sign, from + to − or from − to +; and as many false roots as the number of times two + signs or two − signs are found in succession" (Descartes 1925, p. 160). Maimon solves quadratic equations in his unpublished manuscript on algebra written in Hebrew, *Ma'aseh Choshev* (which is a part of *Hesheq Shelomo*; e.g. p. 186), so it is very likely he was familiar with this kind of analysis.

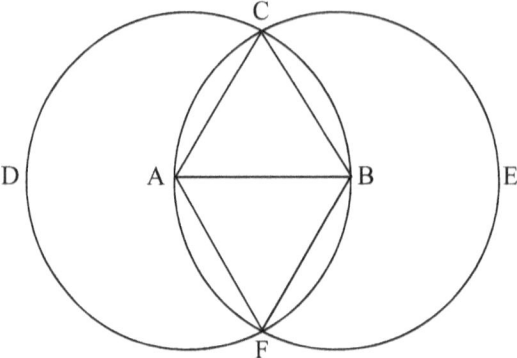

Fig. 15

Although Maimon does not mention it, it is this requirement of presenting both solutions simultaneously that prevents us from applying the method to proofs of theorems. To better understand this requirement and how it applies only to problems and not to theorems, I examine a theorem transformed into a problem. Only once the theorem is transformed into a problem can this method be applied and the requirement of simultaneous presentation of all the cases of the solution be fulfilled. Let us take Euclid's theorem, *Dt.* 90: "If from a given point a straight line be drawn tangent to a circle given in position, the straight line drawn will be given in position and in magnitude" (Taisbak 2003, p. 229). After the general proposition, comes the exposition [*ekthesis*] "For, from the given point *C* let the straight line *CA* have been drawn tangent to the circle *AB* given in position [III.17]", followed by the determination [*diorismos*] "I say that the straight line *CA* is given in position and in magnitude" (Taisbak 2003, p. 229).

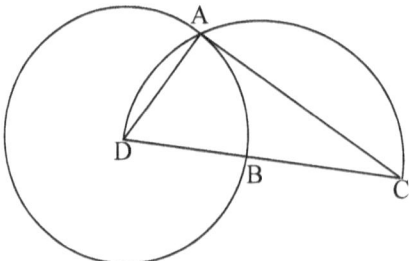

Fig. 16. Based on a diagram from Taisbak 2003, p. 229.

Zheng suggests that *Dt.* 90 should be rephrased such that only the circle and the point C are given, while the tangent AC is not given but sought. He rephrases the proposition in the form of a problem: "Pf.Dt.90: From a *given* point *to draw* a straight line tangent to a *given* circle" (Zheng 2012, p. 106). The presentation of the straight line (tangent to the circle) as sought rather than given is a result of the transformation of the proposition into a problem. For the reason that we cannot ask to solve the problem: "to find the position and magnitude of a tangent", only prove that the given tangent is given in position and magnitude. However, this transformation from a theorem into a problem is informative for understanding the requirement that all the cases of the solution be represented simultaneously. In Euclid's proof, AC is a given straight line which is tangent to a given circle and what is sought is that AC is given in position and magnitude.[194] In Zheng's solution, the straight line AC is constructed, proven to be tangent to the given circle.[195] Thus, using Zheng's solution, we could represent two cases of the solution at the same time: the construction of two straight lines, AC and CG, from a given point C to a given circle BEF. If we insist on adding a second tangent to Euclid's proof, we would need to alter the given exposition and add a new exposition: "For, from the given point C let the straight lines CA and CG have been drawn tangent to the circle AB given in position" and also a new determination: "I say that the straight lines CA and CG are given in position and in magnitude."

---

194 "For, let the center D of the circle have been taken, and let DA, DC have been joined. Since each of D, C is given, therefore [the line] DC is given [Dt. 26]. And angle DAC is right [III.18]; therefore the semicircle described on DC will pass through A [III.31]. Let it have passed, and let it be the [semicircle] DAC; then DAC is given in position [Def. 8]. And the circle AB is also given in position; therefore the [point] A is given [Dt. 25]. But C is also given; therefore AC is given in position and in magnitude [Dt. 26]" (Taisbak 2003, p. 229).
195 Zheng presents the following proof: "Let C be the given point, and BEF the given circle. It is *required to draw* a straight line from the point C touching the circle BEF. Let the center D of the circle be taken, and let DC be joined. Let the semicircle DAC described on the diameter DC, and let it be cutting with the circle BEF at the point A. Let the straight line AC *be joined*, and let DA be joined. *I say that* AC touches the circle EFG. For, since DAC is the angle in the semicircle, DAC is a right angle. So the straight line AC is drawn at right angles to the diameter of the circle BEF from its end the point A, therefore CA touches the circle BEF at A" (Zheng 2012, p. 106).

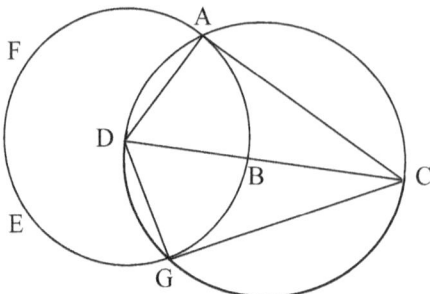

Fig. 17

Hence the requirement of a simultaneous presentation of all the cases of the solution cannot be fulfilled in theorems since it requires changing the exposition and determination, a practice that is not part of Euclidean geometry. Furthermore, presenting all the cases of a solution is not a part of the practice of Euclidean geometry. Greek mathematicians presented only the most difficult case of a solution or a proof, assuming that once presented with the solution of the most difficult case, the readers would be able to arrive at all the possible cases of a solution by themselves.

4.1.6 Analysis of the Various Ways in which a Problem Can be Solved or a Proposition Proven

The sixth kind of analysis is not a technical tool but a general rule. It does not offer a way to arrive at a new proposition to be used as a premise, it only suggests looking for other solutions to given problems (or proofs to given propositions). According to Maimon, the number of ways to solve a problem or prove a proposition is indefinite. Therefore, this method serves as an "inexhaustible source of invention" (*Ueber den Gebrauch*, p. 31; *Das Genie*, p. 380–381).[196] He states that since truths are connected to each other in various ways, propositions proven in a certain way can be proven in other ways as well.[197] Thus insight into the connections between the truths becomes more and more complete. Of all the

---

196 Bogomolny's presentation of 118 different proofs of the Pythagorean Theorem is evidence of how deep the well of productiveness can be (A. Bogomolny, Pythagorean Theorem and its many proofs *from Interactive Mathematics Miscellany and Puzzles*, http://www.cut-the-knot.org/pythagoras/index.shtml, accessed 24 May 2016).

197 This idea that all human knowledge is interconnected is of course not original to Maimon. It appears, for instance, in Descartes' *Discourse on the Method* (*CSM*, Vol. 1, p. 19).

various commentators of Euclid who have proposed alternative solutions and proofs, whose work was familiar to Maimon (such as Clavius and Simson), he mentions only Proclus (*Ueber den Gebrauch*, p. 32), but does not give a specific example. In many cases, an alternative solution or proof can arise when a condition is changed, or a new case of the problem added. Hence in conducting the sixth kind of analysis we can assist in other kinds of analysis. This is not mentioned by Maimon but implied by his theory and accords with his line of thought. For instance, Heron's alternative proof of *Elements* III.20 is based on introducing a new case of the problem (Heath 1956, Vol. II, p. 47–48; see Section 4.1.3). Another example is Proclus' alternative proof of *Elements* I.5: it is based on treating what seems to be a necessary condition in the original proposition as if it is only a pseudo-condition. He offers a solution that does not take into account the production of the equal straight lines further and the proof of equality of the angles under the base, as it appears in Euclid's original proof (Proclus 1789, p. 52–53). The sixth kind of analysis is similar in spirit to another known method for increasing the possibility of invention used by ancient Greek geometers: using one proof (with slight variations) in different solutions. The more geometers repeated using proofs to different situations, one slightly different than the other, or offered a new angle on a known proof, the better they became at inventing new solutions to new problems. Bernard mentions that this also affects the manner in which they do and write geometry since they do not present all the steps but leave some things to be investigated by the reader learning the proof (Bernard 2010, p. 74). Although Maimon suggests presenting various proofs of one case and the Greek geometers suggest using one proof in various cases, both methods encourage searching for as many connections between truths as possible.

Striving to connect truths in various ways is not only a general rule of invention, but also the purpose of human intellectual activity. In his introduction to *Giv'at Hamore*, Maimon describes arriving at perfection not as the attainment of truths but of the connections between truths, an action of the understanding (*Giv'at Hamore*, p. 4–5).[198] Expanding our knowledge can be achieved in varied ways, usually involving not only reason and understanding, but also sensibility. The emphasis put on the latter by Maimon is different in *Giv'at Hamore* and in his works on invention. In *Giv'at Hamore*, the focus is on logical and metaphys-

---

[198] This idea, that the highest degree of perfection of the human understanding is the perfection of the intellectual activity gained from the study of metaphysics, is derived directly from Maimonides' last chapter of *Guide of the Perplexed* (Maimonides 2002, III.54, p. 673). *Giv'at Hamore* is Maimon's commentary to the first book of the *Guide of the Perplexed* and the introduction is mainly dedicated to the final chapter of Maimonides' work.

ical rules of the understanding (such as the principle of contradiction and the concept of cause) referring mostly to undetermined objects. When writing on perfecting the inventive faculty, by contrast, Maimon concentrates more on rules of mathematical analysis, regarding determined objects. In *Giv'at Hamore*, Maimon argues that while the various sciences differ in their subject matter, they all comply with the rules of logic and metaphysics. These rules are the form shared by all fields of knowledge [צורת החכמות] (*Giv'at Hamore*, p. 1–2). Asserting that it is more important to learn how one truth can be inferred from another than learning their content, he urges his readers to direct their efforts not at attaining known truths alone, but at learning how these truths were attained by previous sages. In order to accentuate his claim, he states that there is no advantage in asserting that the earth is round over asserting it is flat, when these propositions are considered by themselves. The only advantage in asserting that "the earth is round" is that this proposition has many connections with other propositions (such as the general laws of motion) and that we arrive at these connections by the action of the understanding [שכל].[199] This contrasts with the assertion that the earth is flat, an assertion that is possible, yet unconnected with other propositions and thus not a product of the action of the understanding (*Giv'at Hamore*, p. 5). It is the form, in other words, the logical connections between truths and the number of such logical connections, that is important in this description of knowledge. To illustrate this, he presents a method that is supposed to help us determine whether the content of a given truth is true or not by the number of its connections to other known truths. Nevertheless, our attempt to arrive at perfection of the inventive faculty involves not only the actions of the understanding and reason but also of our faculty of imagination. In *The Genius and the Methodical Inventor*, Maimon states that he examines real inventions and not merely formal ones, consequently dedicating his theory of invention to mathematical and not philosophical invention (*Das Genie*, p. 372).[200] Even more so,

---

199 Maimon's use of the term *understanding* [שכל] does not accord with his definition of the term in *Logic*, and he should have used the term *reason* [תבונה] instead. In *Logic*, he defines the action of the understanding as judging and the action of reason as inferring. Judging is connecting in an immediate way, whereas inferring is connecting in a mediated way (*Logik*, p. 209). Maimon's definition of the action of the understanding as judgement is similar to Kant's: "Now we can reduce all acts of the understanding to judgments, and the *understanding* may therefore be represented as a *faculty of judgment*" (*CpR*, § A69/B94, p. 106). Following these definitions, the judgment of the earth as round or flat is the action of the understanding, whereas it is the action of reason to find whether this judgment can be inferred from, or can infer, other known truths.

200 "Philosophy itself produces merely *formal* inventions, but not *real* ones. It can invent general methods, formulas and systems, but not truths pertaining to determined real objects. Nathis work is dedicated to actual objects of mathematics, not merely real ones. Furthermore, in *Methods of Invention*, he mentions that his theory of invention is based not only on general logic, but also on mathematical analysis, involving construction and not only logical principles (*Erfindungsmethoden*, p. 139–142). In the sixth kind of analysis, when we are urged to find as many proofs or solutions as possible, we are urged not only to find syllogisms (in which the given proposition serves as a conclusion) but also to find connections given in intuition (that are opaque to reason). Although not mentioned by Maimon, proving a proposition in several ways can give us insights into different properties of the given object, properties on which the different proofs are based. This method can offer a possible unknown truth to be discovered – the connection between these two properties. After proving a proposition in different ways, we can put our efforts into the question of whether and how these different proofs are connected.

In order to examine this claim, let us turn to two different proofs of proposition 30 in *Data:* "If from a given point to a straight line given in position a straight line be drawn making a given angle, the line drawn is given in position" (Taisbak 2003, p. 94). The form of the proofs is similar, since both are based on hypothetical inference. The only formal difference is that the first proof is based on *reductio ad absurdum*, whereas the second is a direct proof. The greater difference between the two proofs is found in their matter, not their form. The first proof is based on an examination of whether the given line changes position and the second proof is a direct proof demonstrating why the given line does not change position. Euclid's proof of the proposition (as it appears in Taisbak's translation) is based on the assumption that something given in position can change its position (*metapitptein*), what Taisbak refers to as *metapipt*.[201] For general use, Taisbak defines *metapipt* as "change of position", in contrast to "given in position" (Taisbak 2003, p. 94).[202] The proposition is proven by *reductio ad absurdum:* if we assume that point *A* remains in its position, and that the straight line *AD metapipt* ("changes position") while preserving the magnitude of angle

---

ural science makes *discoveries* at most, but no inventions. Only pure and applied *mathematics* can rightly claim to make *inventions*" (*Das Genie*, p. 372).
201 Euclid uses only the verb *metapitptein*, whereas it is Taisbak who presents the noun *metapipt* to describe this ability to change position. I thank Michael N. Fried for this observation.
202 Taisbak discusses at length the concept of being given in position and the etymology and meaning of *metapipt*. One of its meanings is 'hop away', thus when a point is given in position, it "may not hop around" and it is identified (Taisbak 2003, p. 93). Lines and figures, on the other hand, have positions but "are 'only' given in *magnitude* so that any object *equal* to a given may serve as well" (Taisbak 2003, p. 96).

ADC, then its position is AZ. Consequently, angle ADC is equal to angle AZC, which is impossible, since angle ADC is greater than angle AZC (since angle ADC is an exterior angle of the triangle and angle AZC is an interior angle; see Elements I.16).[203] Therefore, the position of AD will not *metapipt* and AD is given in position (Taisbak 2003, p. 103).[204]

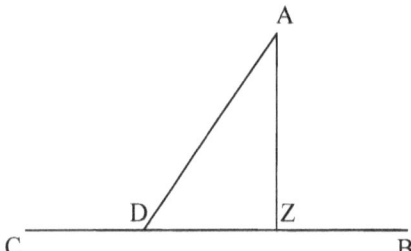

Fig. 18. Based on a diagram from Taisbak 2003, p. 103.

This proposition appears as Proposition 33 in Simson's translation: "If a straight line be drawn from a given point to a straight line given in position, and makes a given angle with it; that straight line is given in position" (Simson 1811, p. 389). His proof is not based on whether the line AD, proven as given in position, does or does not *metapipt*; it uses the previous proposition and the construction of a parallel line (EF) to the given line BC in order to show that AD is given in position. Simson constructs, through the given point A, the line EAF parallel to the straight line BC. BC is given in position therefore EAF is also given in position. Since AD meets the parallels BC and EF, the angle EAD is equal to the given angle ADC, and hence the angle EAD is also given. Simson uses proposition 32 ("If a straight line be drawn to a given point in a straight line given in position, and makes a given angle with it; that straight line is given in position") to conclude that since the line AD is drawn to the given point A, and since point A is on the given straight line EF given in position, and since the line AD makes with the line EF the given angle EAD, then AD is given in position (Simson 1811, p. 389–390).[205]

---

203 Elements I.16: "In any triangle, if one of the sides be produced, the exterior angle is greater than either of the interior and opposite angles" (Heath 1956, Vol. I, p. 279).
204 This proposition has a number of alternative proofs. Taisbak mentions that Menge presents no less than three alternative proofs for it (Taisbak 2003, p. 104).
205 Simson's proposition 32 appears in Taisbak's translation as proposition Dt. 29.

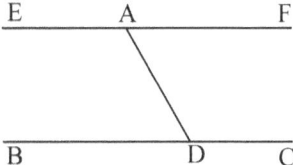

Fig. 19. Based on a diagram from Simson 1811, p. 389.

Since in Simson's proof the two extreme points are given, the line is given in position and does not change position. Other than proving directly what was previously proven indirectly, these two connected proofs can offer new truths to be found. Euclid has established a connection between a line being drawn from a point (*Dt.* 30) and this drawn line not changing position (not *metapipt*). Simson has established a connection between a line being drawn from a given point (*Dt.* 30/Simson's prop. XXXIII) and a line being drawn to a given point (prop. XXXII). Now we can set out to search for a connection between the line not changing position (not *metapipt*) and the givenness in position of a line drawn to a given point (instead of "from a given point").

This method differs from the fifth kind of analysis (analysis of the cases of the solution) since it offers finding new solutions whereas the fifth method is used to arrive at new constructions using one solution. The fifth method is not used to find new solutions but rather only to duplicate a known solution for new cases of the solution, i.e. new constructions. When Maimon mentions a complete solution in the fifth method, he refers to finding all the ways to construct one known solution, not to finding new solutions. Thus, the sixth method can serve as an "infinite source of invention" and at the same time, include solutions which can, each one in itself, be considered as complete. Moreover, the requirement to be able to present all found constructions simultaneously refers to the fifth method wherein all the possible cases of a solution are presented. This requirement is not applied in the sixth method.

If two systems of truths having formal similarities are given, we can attempt to apply a known form of connections of one system to the other. Although not mentioned by Maimon, this idea can be inferred from his method. Thus, for instance, a network of truths in Euclidean geometry can provide new propositions to be determined as sought truths in a non-Euclidean geometry. Using the Pythagorean Theorem as it appears in Euclidean geometry, mathematicians working in the field of hyperbolic geometry (a non-Euclidean geometry) have determined a hyperbolic Pythagorean Theorem as a sought truth to be proven. The assumption of similarities between the two systems is so strong, that even after the publication of an article showing that the hyperbolic Pythagorean Theorem does not exist, mathematicians kept on searching for the proof of such a theorem. Proven by Ungar, the theorem "expresses the square of the hyperbolic length of the hypotenuse of a hyperbolic right angled triangle as a natural "sum" of the squares of the hyperbolic lengths of the other two sides" (Ungar 1999, p. 759). Ungar mentions that the formulation of the hyperbolic Pythagorean Theorem does not have a form analogous to the Euclidean Pythagorean Theorem (Ungar 1999, p. 759). However, it is the assumption of similarities between Euclidean geometry and hyperbolic geometry that ignited the search for a Pythagorean Theorem in this non-Euclidean geometry. This gives new horizons to Maimon's words that "although *Euclid's* proofs are not *means* [of invention], they are *inducements* to the *invention* of new proofs" (*Ueber den Gebrauch*, p. 31–32). In respect of the debate over whether Maimon anticipated non-Euclidean geometries, I believe a contribution can be made by his theory of invention. By and large, commentators have referred to form alone, mentioning how even if we change Euclid's axioms, we could apply the same logical laws to the new system (e.g. Buzaglo 2002, p. 53; Freudenthal 2006, p. 94). But Euclidean geometry can offer the non-Euclidean geometries more than its necessary forms, such as the form of hypothetical inference. It can offer the non-Euclidean geometer a possible sought truth to be found, based on the assumption of some similarities of content between the two systems. Thus, the search for the proof of the hyperbolic Pythagorean Theorem was ignited by determined relations between determined objects, not merely by a general logical form. Indeed, the relation between magnitudes described in the Pythagorean Theorem is in itself form and not matter, but as it is regarded not by itself but rather in relation to determined magnitudes and as determined ratio, it belongs more to the domain of mathematical matter than it does to the domain of logical forms. Nonetheless, the most fundamental aspect for Maimon in exploring truths is their logical connections, which is the subject-matter of the following method.

4.1.7 Logical Analysis

Logical analysis plays an important role in Maimon's theory of invention (discussed also in Chapter 3). In this section I examine his notion of logical analysis as a method. Maimon does not elaborate much on this method, and does not give a specific example from Euclidean geometry.[206] We can extract a definition

---

[206] When describing the kinds of analysis, Maimon does not present a specific example for the

of logical analysis from his quotation of Schwab's definition: "This study of mutual connection or dependency of two propositions or two concepts by means of middle terms or middle concepts is called analysis" (*Ueber den Gebrauch*, p. 22; *Das Genie*, p. 376). Maimon refers to this process as logical analysis (*Ueber den Gebrauch*, p. 22; *Das Genie*, p. 377). In *Methods of Invention*, he defines *logical invention [logische Erfindung]* or *logical analysis [logische Analysis]* as something "according to which we produce various judgments and conclusions by combination" (*Erfindungsmethoden*, p. 140). Although Maimon does not mention this explicitly, there are two methods of logical analysis: regressive analysis and syllogisms. Both methods accord with Schwab's definition adopted by Maimon. Furthermore, Maimon refers to logical analysis as an indispensable condition in any invention (*Ueber den Gebrauch*, p. 22). In this section, I present Maimon's two methods of logical analysis and examine this claim of indispensability. My conclusion is that neither method can serve as a *conditio sine qua non* of any proof or solution. Thus, we should either regard the claim that this method is indispensable as incorrect or assume that by using the term *logical analysis*, Maimon means the principle of contradiction.

In his articles on invention, Maimon does not describe in detail how logical analysis is applied to demonstrations. He only refers to syllogisms in general, without specifying what is entailed in forming them. A thorough description of syllogisms is found in his work on logic. In his articles on invention, Maimon describes logical analysis as presented in Schwab's introduction to the *Data*. There, Maimon refers to logical analysis as the transition from one proposition to another by using intermediate propositions, which he names *regressive analysis* or *analytic method*. However, the method is described in a general manner and does not comply with the rules for forming syllogisms prescribed in *Logic*. Hence, I relate separately to these two descriptions of the methods. It is plausible that Maimon does not give an example of this seventh kind of analysis because he describes the analytic method earlier in the text, using Schwab's work. In order to examine how these methods are applied, I chose to apply both methods on the same example, proposition 1 of the *Data*: "If two magnitudes are given, so is their ratio given" (Schwab 1780, p. 9). This proposition appears as a hypothetical proposition in Schwab's translation of Simson's edition and as a categorical proposition in Simson's own version.[207] The hypothetical proposition would be

---

seventh kind of analysis, but only says in passing: "As a result of all of the above, the already mentioned *logical* analysis" (*Das Genie*, p. 381; *Ueber den Gebrauch*, p. 32).
207 "Wenn zwo Größen gegeben sind, so ist ihre Verhältniß gegeben" (Schwab 1780, p. 9). I have here translated Schwab's German translation of Simson, rather than using Simson's Eng-

used in the first section on regressive analysis, because this method relies on the form of hypothetical judgments. The categorical proposition would be used in the second section, on syllogisms, since Maimon's theory of syllogism is based on judgments in the categorical form.

4.1.7.1 Regressive Analysis

In discussing Schwab's introduction to Euclid's *Data*, Maimon quotes Schwab on how, in order to arrive from proposition A to proposition E, one needs to find the intermediate propositions B, C and D. Once these are found, one can also use them to arrive at proposition A from proposition E (*Ueber den Gebrauch*, p. 21–22). Maimon then remarks:

> [...] this explanation or development of the concept of *analysis* that *Mr. Schwab* has presented concerns only *logical analysis*, which certainly must take place in all investigations as a *conditio sine qua non*; every demonstrative proposition, being a *conclusion*, must be resolved into its *premises*; however, it is far from exhausting the concept of *analysis as such*. (*Ueber den Gebrauch*, p. 22)

Based on Schwab's description of a passage from one proposition to another via intermediate propositions, we can present Euclid's proof of the proposition "If two magnitudes are given, so is their ratio given" (Schwab 1780, p. 9), as follows:
Let us take the following propositions:

(P): "A and B are given in magnitude"
(Q): "C is given in magnitude and equal to A, and D is given in magnitude and equal to B"
(R): "A is to B as C is to D"
(S): "The ratio between A and B is given"

Using these propositions, we can prove proposition 1 in *Data*, if (P) then (S) (if A and B are given in magnitude then the ratio between A and B is given) as follows:

> If (P) then (Q): "If A and B are given in magnitude, then C is given in magnitude and equal to A, and D is given in magnitude and equal to B."
>   * based on *Data*, Def. 1: "Spaces, lines and angles are said to be given in magnitude, when spaces, lines and angles, which are equal to them, can be found" (Schwab 1780, p. 4).
>   If (Q) then (R): "If C is given in magnitude and equal to A, and D is given in magnitude and equal to B, then A is to B as C is to D."

---

lish translation ("The ratio of given magnitudes to one another is given"; Simson 1804, p. 360), since the first is in hypothetical form and the latter in categorical.

If (R) then (S): "If A is to B as C is to D, then the ratio between A and B is given."
  * based on *Data*, Def. 2: "A ratio is said to be given, when a ratio equal to it, of a given magnitude to a given magnitude, can be found" (Schwab 1780, p. 4).

Thus, we can arrive from (P) to (S) by using propositions (Q) and (R), and based on the first two definitions of the *Data*, in a manner similar to Schwab's description of arriving at proposition E from proposition A. A problem arising in this description is the placement of the definitions. The proof is based on definitions I and II in the *Data* and without them we cannot conduct the proof. Yet these definitions do not serve as intermediate propositions in the transition from P to S. Were these definitions used as intermediate propositions, we could not have gone forward with the proof in the required manner, where a conclusion of one hypothetical proposition becomes the premise of the following one. If we had included the definitions, it would have been as follows:

If (P) and (Def. 1) then (Q): "If A and B are given in magnitude, and if spaces, lines and angles are said to be given in magnitude, when spaces, lines and angles, which are equal to them, can be found, then C is given in magnitude and equal to A, and D is given in magnitude and equal to B."

A further difficulty with including the definitions as part of the intermediate propositions is that the definitions are general and the proof is particular (it refers to the magnitudes A, B, C and D). In syllogisms, this problem does not arise because their structure includes a universal and a particular premise.

Let us look into the proof of the opposite direction, "if (S) then (P): "If the ratio between A and B is given, then A and B are given in magnitude". This proposition does not appear in the *Data*, probably because ratio is defined in the *Elements* as something that is between two magnitudes.[208] From Euclid's definition it follows that if a ratio is given, then it is necessary for magnitudes (of the same kind) to be given. The proof of the opposite direction is as follows:

If (S) then (R): "If the ratio between A and B is given, then A is to B as C is to D".
  * following *Data*, Def. 2: "A ratio is said to be given, when a ratio equal to it, of a given magnitude to a given magnitude, can be found" (Schwab 1780, p. 4).
  * It is better to rephrase proposition (R) as (R'): "There is a ratio between the given magnitudes C and D, equal to the given ratio between A and B" so that it follows the definition that a ratio is given when another ratio equal to it is given: If (S) then (R'): "If the ratio between A and B is given, then there is a ratio between the given magnitudes C and D, equal to the given ratio between A and B".

---

208 *Elements*. Book V, Def. III: "Ratio is a mutual relation of two magnitudes of the same kind to one another, in respect to quantity" (Simson 1811, p. 119).

If (R') then (Q): "If there is a ratio between the given magnitudes C and D, equal to the given ratio between A and B, then C is given in magnitude and equal to A, and D is given in magnitude and equal to B".

* Here we encounter another problem: in order to arrive at the sought givenness of the magnitudes from the given ratio, we would need to use Def. 3 of Book V of the *Elements*. But then, if we were to use this definition now, we might as well have used it at the beginning of the proof, making the entire inference from (S) to (P) redundant. Were we not to use this definition, we would have no basis for arriving from a given ratio to given magnitudes. Furthermore, if we use the given ratio between C and D to infer the givenness of the magnitudes C and D, we can in the same way also infer from the given ratio between A and B that A and B are given in magnitude and thus make the inference from (Q) to (P) redundant:

If (Q) then (P): "If C is given in magnitude and equal to A, and D is given in magnitude and equal to B, then A and B are given in magnitude".

* based on *Data*, Def. 1: "Spaces, lines and angles are said to be given in magnitude, when spaces, lines and angles, which are equal to them, can be found" (Schwab 1780, p. 4).

It seems that regressive analysis, the transition from one proposition to another by the use of intermediate propositions, cannot be fully applied to *Dt.* 1. It cannot, therefore, serve as a method indispensable to every proof or solution. We should either conclude that Maimon was mistaken claiming that this method is indispensable to any proof or change his notion of logical analysis to mean no more than the use of the principle of contradiction as a necessary condition of any proof.[209]

4.1.7.2 Syllogisms

As discussed in the previous chapter, Maimon's general notion of invention is based on syllogism. He repeatedly states that in order to invent, we need to find propositions that can serve as premises or conclusions (see *Ueber den Gebrauch*, p. 2; p. 12–13; *Das Genie*, p. 366–367; *Erfindungsmethoden*, p. 140). Maimon also argues that syllogisms do not provide us with the tools to find the sought premises and hence their use is insufficient for invention (*Das Genie*, p. 377; *Ueber den Gebrauch*, p. 22–23). Testimony to the importance of logical analysis is found in Maimon's fourth kind of analysis, analysis of the object. This method of using construction, involving changing the object given in intuition, aims at finding sought premises that will serve as intermediate premises and thus take part in logical analysis (*Ueber den Gebrauch*, p. 29).

---

[209] I refer here to the principle of contradiction since it is the principle usually mentioned by Maimon (e.g. *Erfindungsmethoden*, p. 141; *Logik*, p. 19), but the argument applies to the principle of identity as well. Both are principles of general logic and are necessary conditions to any science (*Logik*, p. 4).

Syllogisms are defined by Maimon as "nothing but a means, to derive *new knowledge* from *given knowledge*, according to *a priori forms* established in the faculty of cognition" (*Ueber den Gebrauch*, p. 1–2). Though Maimon defines syllogisms as a method of inventing new knowledge, he does not, in his works on invention, elaborate what constructing such syllogisms entails. In *Logic*, however, he does present such specification. Among the rules for forming syllogisms are, for instance, the following:

- a conclusion cannot have more or less than three different terms (*Logik*, p. 89)
- the middle term is not allowed to be particular in both premises (*Logik*, p. 91)
- the major premise has to be universal (*Logik*, p. 100)
- if the minor premise is particular, so is the conclusion particular (*Logik*, p. 100)

In his articles on invention, Maimon does not mention whether by syllogisms he is referring to his notion of syllogisms as based on the principle of determinability, as presented in *Logic*, or as it appears in traditional logic. In *Logic*, Maimon states that "all forms of inference are based on the principle of determinability" (*Logik*, p. 104). He rephrases the inference "if *a* is *b*, and *b* is *c*, then *a* is also *c*", according to relation of determinability: "*c* is the determinable in *b*, and *b* the determinable in *a*, therefore *c* is also the determinable in *a*", adding that *c* is attributed to *a* in a mediated way (*Logik*, p. 89). However, the application of the principle of determinability to the form of hypothetical inference is problematic. Let us consider the following example, mentioned by Bergman: 'Take the determinable "figure" which is determined as "triangle", and we determine the "triangle" by means of a subsequent determinant as "equilateral." We have determined not only the "triangle" but also the determinable "figure." The equilateral triangle is also an equilateral figure' (Bergman 1967, p. 280). Using Maimon's notation presented above: *figure* (*c*) is the determinable of *triangle* (*b*), *triangle* (*b*) is the determinable of *equilateral* (*a*), therefore *figure* (*c*) is determinable of *equilateral* (*a*) in a mediate way, by using *triangle* (*b*). A problem arises since we can replace *triangle* with *quadrilateral*: *figure* (*c*) is the determinable of *quadrilateral* (*d*), *quadrilateral* (*d*) is the determinable of *equilateral* (*a*), therefore *figure* (*c*) is determinable of *equilateral* (*a*) in a mediate way, by using *quadrilateral* (*d*). In this case, *c* is attributed to *a* in a mediated way (as Maimon describes in *Logik*, p. 89), but the middle term has changed from *b* (*triangle*) to *d* (*quadrilateral*). This is problematic since according to Maimon, "a predicate cannot have several subjects even disjunctively" (*Tr.*, p. 380) whereas a subject can have several predicates disjunctively. This means that *equilateral* should have only one determinable, either *triangle* or *quadrilateral*. As stated by Maimon, in a synthesis which is an absolute concept, a subject is the part that can be thought without the other part (i.e. it can be thought by itself or in another synthesis). The predicate is the part that cannot be thought without reference to the other part of the synthesis. He presents the example of being right- or oblique-angled: a triangle can be thought in itself or in other synthesis whereas being right- or oblique-angled cannot be thought without the triangle (*Tr.*, p. 84–85). In my earlier discussion in Section 3.6 on the principle of determinability, I have shown that *right-angled* can be thought without thinking *triangle* and therefore *triangle* is not its determinable. A question arises: why does *right-angled* have a different determinable than *equiangular*, being that both are angles? Furthermore, Maimon claims that *equilateral* is a determinant of *side*, not of *triangle*, and that *right-angled* is a determinant of *angle* (*Logik*, p. 188). Consequently, another question arises: is *triangle* the determinable of *equilateral* or is it *side*? Furthermore, the assertion that the determinable of *right-angled* is *angle* does not correspond with the assertion that the determinable is *triangle*. These issues are left unaccounted for by Maimon. Moreover, in Euclid's *Elements* we find that *rectilineal figure* can be determined as *trilateral* or *quadrilateral* (Def. 21; Heath 1956, Vol. I, p. 154). *Triangle* can be determined as *equilateral* and as *right-angled* (Def. 20 and Def. 21), but then again, so can *quadrilateral* be determined as *equilateral* and *right-angled* (Def. 22). Thus, *equilateral* and *right-angled* have both *triangle* and *quadrilateral* as their determinables. This does not accord with Maimon's own requirement that a determination will have one determinable and it conflicts with his assertion that *right-angled* cannot be thought without *triangle*. Were we to suggest that *quadrilateral* is a composition of triangles and thus *triangle* serves as a middle term to connect *quadrilateral* with *equilateral*, we would have encountered a difficulty because *trilateral* and *quadrilateral* are disjunctive determinations of *rectilineal figure*. In defining *quadrilateral* as "a figure enclosing two triangles", we are presented with one of the following options: first, *figure* (*a*) is the determinable of *triangle* (*b*), *triangle* the determinable of *quadrilateral* (*c*) and *quadrilateral* is the determinable of *equilateral* (*d*). In this case, a problem arises with the introduction of the determination *equilateral triangle*, where *triangle* is the determinable of *equilateral* and not of *quadrilateral*. The second option is that *figure* (*a*) is the determinable of *triangle* (*b*), *triangle* is the determinable of *equilateral* (*c*) and *equilateral* is the determinable of *quadrilateral* (*d*). In this case, a problem arises when we introduce the *rectangle* – a quadrilateral figure that is not equilateral. Due to all the mentioned difficulties in presenting the application of the principle of determinability, let us turn to syllogisms as they appear in traditional logic.

Following Maimon's rules for syllogisms, let us turn again to our example, *Dt.* 1: "The ratio of given magnitudes to one another is given" (Simson 1804,

p. 360). The rule that states that the major premise has to be universal (*Logik*, p. 100) can be applied on *Dt.* 1. We could form the following syllogism using Def. I as the universal proposition, which is the major premise. Since the definition, "Spaces, lines and angles are said to be given in magnitude, when equals to them can be found" (*Data*, Def. I: Simson 1811, p. 367), is in the hypothetical form, we need to adjust it in order to use it as a major premise:

Major premise: "Lines given in magnitude are lines that have their equals"
Minor premise: "Lines A and B are lines given in magnitude"
Conclusion: "Lines A and B are lines that have their equals"

The major premise is universal and the minor premise is a particular instance. Harari points out that while mathematics applies general theorems to a particular case, syllogistic inference only deals with the subsumption of a particular case under a general principle (Harari 2004, p. 95).[210] She concludes that "Aristotle's theory of demonstration cannot be made to accommodate Greek mathematics" (Harari 2004, p. 90). In order to form a syllogism based on the proof of *Dt.* 1, the use of a particular instance here was unavoidable, yet it does not accord with the demand that the minor premise be a particular case of the major premise.

This syllogism accords with another rule presented by Maimon – having three terms: (a) lines A and B; (b) lines given in magnitudes; (c) lines that have their equals; the middle term being "lines given in magnitude". However, this syllogism does not express the relation of determinability Maimon ascribes to syllogisms: a major premise is any premise in which the major term is the determinable of the middle term, and a minor premise is any premise in which the middle term is the determinable of the minor term (*Logik*, p. 91). We cannot apply such relation of determinability between the major premise, the minor premise and the middle term as presented in the syllogism based on *Dt.* 1.

---

[210] Accordingly, in a syllogism formed by Alexander of Aphrodisias, based on a part of the proof of *Elements* I.1, the major premise is a common notion (a general principle), and the minor premise is a particular premise regarding determined geometrical magnitudes:
"(1) Things equal to the same thing are also equal to each other;
(2) Sides CA and CB of the triangle are things equal to the same thing, i.e., to side AB;
Therefore, (3) CA and CB are equal to each other" (Harari 2004, p. 94–95). As Harari mentions, the equality of the sides CA and CB is only assumed in the second premise of Alexander's inference while in *Elements* I.1, this equality is demonstrated by using construction (Harari 2004, p. 94–95). Alexander's first premise is a general principle, while Euclid's theorem is a general theorem: a general principle is an immediate truth, recognized as such without the presentation of a necessary proof (e.g. "Grundsatz", in: Wolff 1734, p. 602), whereas a general theorem used in mathematical inference is demonstrated using spatial relations.

In conclusion, the two methods of logical analysis, regressive analysis and syllogisms, cannot be taken to be indispensable conditions for any proof. Thus, we can either assume that Maimon was mistaken in making such a claim or that he meant to claim that it is the use of the principle of contradiction that is necessary in any proof and invention, erroneously using the term *logical analysis* instead of *the principle of contradiction*.

4.2 Conversion

The method of conversion has multiple functions. It is used not solely for inventing new propositions (the converse of given propositions), but also for determining whether a proposition is derived or original, analytic or synthetic. Moreover, by determining whether a proposition is derived or original, the method of conversion plays an important part in the method of generalization (discussed in Section 4.3).

Conversion was a well-known logical and geometrical tool in Maimon's time.[211] Logical conversion is defined as interchanging the subject and its attribute: the subject in a given proposition becomes the attribute in its converse proposition and the attribute in the given proposition becomes the subject of the converse, without changing the truth-value of the proposition (unknown, "Proposition [*Logique*]", in: *Encyclopédie*, Vol. 13, p. 479). In geometrical conversion we interchange the given and the sought: we assume the given proposition as the sought conclusion of the converse, and the sought proposition as the given proposition of the converse (d'Alembert, "Converse [*en Géométrie*]", in: *Encyclopédie*, Vol. 4, p. 166). Perhaps since these methods of logical and geometrical conversions were already well-known, Maimon only mentions conversion as a method of invention without elaborating further on how to apply it. For instance, if we are given a proposition (such as *Elements* I.24) and we wish to find its converse (assuming there is one; in this case, *Elements* I.25), we will not find tools to conduct the conversion and proof in Maimon's work on invention.[212] Maimon could have offered the reader rules of logical conversion, such as

---

[211] A *Converse* is also called *inverse* (d'Alembert, "Converse [*en Géométrie*]", in: *Encyclopédie*, Vol. 4, p. 166).

[212] *Elements* I.24, "If two triangles have the two sides equal to two sides respectively, but have the one of the angles contained by the equal straight lines greater than the other, they will also have the base greater than the base" (Heath 1956, Vol. I, p. 296).

the following rule offered by Aristotle: "if *A* holds of every *B*, then *B* will hold of some *A*."[213] He could have also offered the methodical inventor tools for geometrical conversions such as *reductio ad absurdum*, the method most used by Euclid when proving converse propositions.[214]

The distinction between logical and geometrical conversion should be based not only on whether the two interchanged terms are the subject and predicate of a proposition (logical conversion) or two propositions (given and sought – as in geometrical conversions). I suggest that the distinction could also be based on whether the interchanged terms are complete or partial. In logical conversion we interchange the entire subject with the predicate. For instance, "all humans are mortal" is converted into: "some mortals are humans" (with the required change of quantifier). In geometrical conversion, on the other hand, the subject remains the same and what changes are its predicates. For instance, the triangle as a subject remains the same in both *Elements* I.5 and its converse *Elements* I.6.[215] The interchange in the two propositions is whether the triangle is isosceles and has two equal angles at the base. The given in *Elements* I.5 is an isosceles triangle and it is this equality of the sides that becomes the sought in *Elements* I.6. The triangle itself remains as given in both the original proposition and its converse. Maimon does not give an account of the difference in the definitions of the two types of conversion as described above. However, he could have presented it when discussing the example based on *Elements* I.30, where he describes how only a part of the subject is interchanged with the predicate. The

---

Elements I.25: "If two triangles have the two sides equal to two sides respectively, but have the base greater than the base, they will also have the one of the angles contained by the equal straight lines greater than the other" (Heath 1956, Vol. I, p. 299).

213 This rule of conversion is true for both universal and particular affirmatives: "<If A holds of every B, then B will hold of some A. For if of none, A will hold of no B. But it was supposed to hold of every B. – Similarly if the proposition is particular. For if A holds of some B, then it is necessary for B to hold of some A. For if of none, then A will hold of no B>" (Aristotle, *Prior Analytics*, 1.2, 25a17–22, from: Alexander of Aphrodisias 1991, p. 90).

Socher comments that Maimon was familiar with some of Alexander of Aphrodisias' work via the works of Maimonides and that Maimon mentions Alexander in *Hesheq Shlomo*, folio 297 (Socher 2006, p. 187).

214 On Euclid's use of the method of *reductio ad absurdum* in proving conversion, see Proclus 1789, p. 115.

215 *Elements* I.5: "In isosceles triangles the angles at the base are equal to one another, and if the equal straight lines be produced further, the angles under the base will be equal to one another" (Heath 1956, Vol. I, p. 251).

Elements I.6: "If in a triangle two angles be equal to one another, the sides which subtend the equal angles will also be equal to one another" (Heath 1956, Vol. I, p. 255).

Only the first half of *Elements* I.5 is converted in *Elements* I.6.

given proposition is: "if *AB* is parallel to *CD*, and *EF* is parallel to *CD*, then *AB* is parallel to *EF*." This proposition, he continues, is synthetic and convertible: *CD* is parallel to *EF*, which Maimon describes as "a part of the subject" and this part is unchanged in the process of conversion. What interchanges is what Maimon calls another part of the subject (*AB* is parallel to *CD*) with another part of the predicate (*AB* is parallel to *EF*) (*KrU*, p. 369–370). Maimon mentions the hypothetical form of inference that geometrical propositions are based on ("if *a* is *b* and *b* is *c*, then *a* is *c*"). In this form of inference the subject (*a*) appears both as a part of the given ("*a* is *b*") and in the sought ("*a* is *c*"). Unlike the logical conversion mentioned above, in geometrical conversion there is no change of quantifier. Thus, the criterion of whether the subject in its entirety interchanges in the process of conversion can help us differentiate between logical and geometrical conversion.

While the method of conversion was not invented by Maimon, he did invent a new use for it: it can serve as a criterion for determining whether a proposition is derived or original, analytic or synthetic. First, Maimon uses the ability (or inability) to convert a proposition as a criterion for determining whether this proposition is derived or not. In his articles on invention from 1795, Maimon claims that if we cannot convert a proposition, then it is not an original proposition. If we can convert the proposition, it can be either original or derived, since "all *original* propositions of *mathematics* are convertible and only the ones which are *derived* from them are not always [convertible]" (*Ueber den Gebrauch*, p. 33–34; *Das Genie*, p. 382). Maimon does not present an example of a proposition that is both derived and convertible, but we can offer the Pythagorean theorem as such an example.[216]

At times, Maimon refers to derived propositions as corollaries. His use of the term is in the meaning of inference, which was common in his time. In Euclidean geometry, however, a corollary is considered to be a product of chance.[217] Mai-

---

216 The Pythagorean theorem, *Elements* I.47 ("In right-angled triangles the square on the side subtending the right angle is equal to the squares on the sides containing the right angle;" Heath 1956, Vol. I, p. 349), is convertible and its converse is *Elements* I.48 ("If in a triangle the square on one of the sides be equal to the squares on the remaining two sides of the triangle, the angle contained by the remaining two sides of the triangle is right;" Heath 1956, Vol. I, p. 368). *Elements* I.47 is also a derived proposition since Pappus presented a more general version of the proposition, from which it is derived: "If *ABC* be a triangle, and any parallelograms whatever *ABED*, *BCFG* be described on *AB*, *BC* and if *DE*, *FG* be produced to *H*, and *HB* be joined, the parallelograms *ABED*, *BCFG* are equal to the parallelogram contained by *AC*, *HB* in an angle which in an angle which is equal to the sum of the angles *BAC*, *DHB*" (Heath 1956, Vol. I, p. 366).
217 For the notion of corollary as inference see: "Folgern", in: Adelung 1811, Band 2, p. 239–240.

mon uses the term *corollary* when presenting *Elements* I.16 as a proposition that can be derived from a more general proposition.[218] He refers to it as a corollary of his new universal proposition (*Ueber den Gebrauch*, p. 34; *Das Genie*, p. 383). Similarly, he regards the proposition about the congruence of two triangles (*Elements* I.4) as a corollary of the proposition demonstrating similarity between two given triangles (*Elements* VI.4).[219] Maimon refers to the equality of the sides of the triangles as a determination of the subject (*Ueber den Gebrauch*, p. 35). What is being determined is the ratio between each pair of sides that is being compared in the proportion that characterizes pairs of similar triangles. In the case of congruence of a pair of given triangles, this ratio is one. Maimon writes that *Elements* I.4 is "a mere corollary [*Corollar*]" (*Ueber den Gebrauch*, p. 35). Due to the use of the method of superposition, many mathematicians refer to *Elements* I.4 as a definition or an axiom rather than a corollary.[220] It is plausible that Maimon refers to this proposition as derived and not original because it is not convertible.[221]

Initially, Maimon viewed conversion only as a criterion for differentiating between original and derived propositions: If a proposition cannot be converted, then it is derived and if it can be converted, then it is either derived or original. However, he later added that conversion can also be used as a criterion for differentiating between synthetic and analytic propositions. In the "Concluding re-

---

Proclus defines a corollary as "a theorem, unexpectedly emerging from the demonstration of another problem, or theorem. For we seem to fall upon corollaries, as it were, by a certain chance; and they offer themselves to our inspection, without being proposed, or investigated by us" (Proclus 1789, p. 100).

218 *Elements* I.16: "In any triangle, if one of the sides be produces, the exterior angle is greater than either of the interior and opposite angles" (Heath 1956, Vol. 1, p. 279). Maimon presents a proposition that is more general: "If a side of a straight-lined figure is extended, then the exterior angle is greater than any interior opposite angle, taken together with the adjacent angle of the exterior angle amounts to less than two right angles" (*Ueber den Gebrauch*, p. 33).

219 *Elements* VI.4: "In equiangular triangles the sides about the equal angles are proportional, and those are corresponding sides which subtend the equal angles" (Heath 1956, Vol. II, p. 200) is a proposition regarding similarities between two given triangles. It serves Maimon as a universal proposition for the proposition regarding congruence between two given triangles, *Elements* I.4: "If two triangles have the two sides equal to two sides respectively, and have the angles contained by the equal straight lines equal, they will also have the base equal to the base, the triangle will be equal to the triangle, and the remaining angles will be equal to the remaining angles respectively, namely those which the equal sides subtend" (Heath 1956, Vol. I: 247).

220 The first suggested by Maimon's predecessor Peletier and the latter suggested by Russell and conducted by Hilbert (Heath 1956, Vol. I, p. 249).

221 This proposition appears in the *Encyclopédie* as an example of a proposition that cannot be converted (d'Alembert, "Converse [*en Géométrie*]", in: *Encyclopédie*, Vol. 4, p. 166).

mark" [*Schlußanmerkung*] of *Critical Investigations of the Human Spirit* (*Kritische Untersuchungen über den Menschlichen Geist*; 1797), he describes how conversion can be used as a method of invention. Maimon refers to this method as "a very *important discovery*" (*KrU*, p. 361), asserting that we can determine whether a proposition is analytic or synthetic based on our ability or inability to convert it. He explains that all original geometrical propositions are synthetic and can be converted without any alteration of quantity because they are of the form "all *A* are *B*". He also asserts that propositions of the kind "all *A* are *B*" can be fully converted without altering the quantity: "all *B* are *A*". The connection between subject and predicate is not in the concept but only in the object. In analytic propositions, on the other hand, a change of quantifier is needed. The predicate is connected to the subject in the concept (the predicate is more general than the subject). Since analytic propositions are of the form "all *AB* are *A*," they cannot be converted without changing the quantifier. The conversion of "all *AB* are *A*" is not "all *A* are *AB*" (which is false), but "some *A* are *AB*" (*KrU*, p. 361–362). Two difficulties arise from this argument: Maimon cites the logical rule of conversion for the form "all *A* are *B*" incorrectly and he mistakenly confuses logical conversion with geometric conversion. After presenting the axiom "the straight line is the shortest between two points" as an original synthetic proposition that can be converted without alteration of quantity, Maimon adds that *Elements* I.5 is the same kind of conversion (*KrU*, p. 362).[222] He asserts that both propositions are of the form "all *A* are *B*" and that propositions in this form can be converted, without a change of quantifier, into "all *B* are *A*."[223] However, Aristotle proved that the logical conversion of "all *A* are *B*" is not "all *B* are *A*" but "some *B* are *A*." Maimon is inaccurate not only in citing the logical rule itself, but also in treating the conversion of *Elements* I.5 as a logical conversion and not as a geometric one.[224] As Heath remarks, a logical conversion of *El*-

---

[222] "[...] da hingegen dieser Satz: alle *a* sind *b*, nicht darum wahr ist, weil *b* in *a* enthalten, sondern weil beide im *Objekte* unzertrennlich sind. Es ist daher auch umgekehrt wahr, daß *alle b* sind *a* [...] Die Axiome, z.B. die gerade Linie ist die kürzste zwischen zweien Punkten, läßt sich unabgeändert umkehren. Eben so auch Euklides I. B. 5 prop." (*KrU*, p. 362)
[223] According to Freudenthal, Maimon mistakenly claims that by proving that in isosceles triangles the angles at the base equal one another makes proving its converse redundant. However, Freudenthal remarks, there is a need to prove the inverse proposition, that all triangles which are not isosceles triangles do not have equal angles at their bases (Freudenthal 2006, p. 98).
[224] Heath remarks that *Elements* I.6 is only a geometric conversion of *Elements* I.5 (Heath 1956, Vol. 1, p. 256). In his commentary on *Elements* I.6, Heath also presents De Morgans' proof of how this proposition can be logically deduced from *Elements* I.5 and *Elements* I.8 taken together (Heath 1956, Vol. 1, p. 256). De Morgan's proof, alongside Heath's correction that the logical conversion of *Elements* I.6 involves the change of quantifier, make *Elements* I.6 a good confirmation

ements I.5 would be "*some* triangles with two angles equal are isosceles" (Heath 1956, Vol. 1, p. 256). Maimon's treatment of the conversion of *Elements* I.5 as a logical conversion rather than geometrical is evident from his application of a logical rule on it and the use of the definition of conversion as an interchange of subject and attribute, i.e. a logical conversion.[225] Were he to conduct a geometrical conversion, he would not have described the rule as an interchange between subject and attribute but as an interchange between a given proposition and a sought conclusion, whereas the subject remains in the same position in the converse proposition. What is interchanged in *Elements* I.5 and *Elements* I.6 is not a subject and an attribute, but a given proposition ("Two sides in a triangle are equal to one another") and a sought conclusion ("two angles in a triangle, subtend two given sides, are equal"). We can describe the difference between logical and geometrical conversions also in terms of the difference between judgment (connecting subject and predicate immediately) and inference (connecting subjects and predicates in an indirect manner), that is, between the action of the faculty of understanding and the faculty of reason.

4.3 Generalization

This method is aimed at finding a proposition which is more general than the given one. For Maimon, the importance of this method lies in the fact that that the more general a proposition is, the more often we can use it as a premise in a syllogism (*Ueber den Gebrauch*, p. 32). According to Maimon, propositions are not always formulated in their highest universality and yet mathematicians scarcely use this method. His example for the application of this method is based on *Elements* I.16: "If a side of a triangle is extended, so the exterior angle is greater than either of the interior and opposite angles" (*Ueber den Gebrauch*, p. 32–33).[226] First he mentions Clavius' claim that this proposition cannot be converted. The converse, "if the exterior angle (a figure, whose side is extended) is greater than either interior and opposite angles, so is the figure a

---

of Maimon's assertion that propositions converted with a change of quantifier are analytic and derived.

225 I am following the definitions of logical and geometrical conversions as they appear in the *Encyclopédie*, and cited above (unknown, "Proposition [*Logique*]," in: *Encyclopédie*, Vol. 13, p. 479; d'Alembert, "Converse [*en Géométrie*]," *Encyclopédie*, Vol. 4, p. 166).

226 Maimon's translation has a similar meaning to Heath's translation: "In any triangle, if one of the sides be produced, the exterior angle is greater than either of the interior and opposite angles" (Heath 1956, Vol. 1, p. 279).

triangle" is incorrect since the first part of this proposition is also true for a quadrilateral figure (*Ueber den Gebrauch*, p. 33).[227] Maimon offers a generalization of this proposition, which he refers to as being in its highest universality: "If a side of a straight-line figure is extended, then the exterior angle is greater than any interior opposite angle, which taken together with the adjacent angle of the exterior angle amounts to less than two right angles" (*Ueber den Gebrauch*, p. 33). But it seems that Maimon has taken a wrong turn with Clavius' example. First of all, Maimon adds a condition ("the sum of any interior opposite angle taken together with the adjacent angle of an exterior angle amounts to less than two right angles") which is not a part of the original proposition (*Elements* I.16) but a variation of a condition in *Elements* I.32. This variation is not original and, as Heath comments, it appears in Proclus' commentary of *Elements* I.16.[228] Second, Clavius rightly points out that the converse of *Elements* I.16 is incorrect since we can find a quadrilateral that can replace the triangle in the proposition. But Maimon takes this possibility and attempts to attribute it to any quadrilateral form, and even further, to any straight-lined form. Even though there are straight-lined forms that fulfill the condition: "if a side of the given figure is extended, then the exterior angle is greater than any interior opposite angle", not all of them do. For instance, an isosceles trapezium *ABCD*: Assuming the obtuse angles, angles *DAB* and *ABC*, are 110 degrees each, it follows that the interior angles, angles *BCD* and *CDA*, are 70 degrees each. If we prolong side *AB*, the exterior angle *EBC* equals 70 degrees. In this case, the exterior angle *EBC* does not fulfill the condition that it should be greater than any interior opposite angle (as it is equal to angles *BCD* and *CDA*). Also, the second condition mentioned by Maimon, that "any one of the interior opposite angles taken together with the adjacent angle of the exterior angle amounts to less than two right angles," is not fulfilled: the opposite interior angle *BCD* taken together with the adjacent angle of the exterior angle *ABC* amount to two right angles, not less.

---

227 Clavius was not the first to comment that this proposition cannot be converted since it is also true for quadrilaterals. We find the same claim made by Philoponus in his commentary on *Posterior Analytics*, when discussing Aristotle's definition of the universal (Philoponus 2008, p. 76).
228 *Elements* I.32: "In any triangle, if one of the sides be produced, the exterior angle is equal to the two interior and opposite angles, and the three interior angles of the triangle are equal to two right angles" (Heath 1956, Vol. I, p. 316). Proclus' new enunciation of *Elements* I.16 is combined with a variation of *Elements* I.32: "In any triangle, if one side be produced, the exterior angle of the triangle is greater than either of the interior and opposite angles, and any two of the interior angles are less than two right angles" (Heath 1956, Vol. I, p. 280).

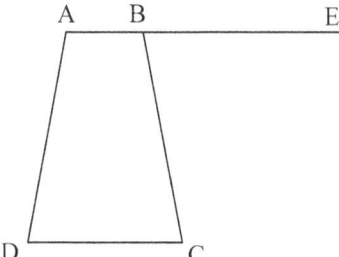

Fig. 20

It seems that Clavius' claim, that *some* quadrilaterals fulfill the first condition, was changed by Maimon into *all* quadrilaterals fulfilling the first condition and was even extended to include other straight-lined forms (such as hexagons and octagons). But as the example of the trapezium shows, there are cases in which Maimon's new proposition is false. Nevertheless, this is not to say that the method itself is redundant. An example of the productiveness of this method is Pappus' extension of the Pythagorean theorem which can be applied not only to right-angled triangles but to any triangle in general.

Maimon presents the converse of his new proposition in order to demonstrate that the new proposition is given in its highest universality, as an original proposition: "If the exterior angle (of a straight-lined figure whose side is extended) is greater than any interior opposite angle, then this [interior opposite angle] with the adjacent angle of the exterior angle are less than two right angles" (*Ueber den Gebrauch*, p. 33). As mentioned in the previous section, he suggests a tool for identifying whether the given proposition is derived or original. If it is derived, then a more general proposition can be found. Therefore, the means for determining whether a proposition is derived or original consists of trying to convert it. According to Maimon, all original mathematical propositions are convertible. If a proposition can be converted, then it may be original or derived. If it cannot be converted, then the proposition is derived. If a proposition is derived and we do not have the proposition that it is derived from, then there is a lacuna to be filled (*Ueber den Gebrauch*, p. 33–34; *Das Genie*, p. 382). Maimon presents the following example of how the inability to convert a proposition marks that the method of generalization can be applied. The proposition: "If in two triangles the sides taken separately are equal to each other, then also their opposite angles taken separately are equal to each other" (*Ueber den Gebrauch*, p. 34) is not convertible.[229] "This is a secure indicator," writes Maimon, "that there is a *gap* here,

---

229 This proposition is similar to *Elements* I.4: "If two triangles have the two sides equal to two

which must be filled by a *more general* proposition" (*Ueber den Gebrauch*, p. 34). Thus, by using the method of generalization, we can arrive at the more general proposition (i.e. from congruence to similarity): "If the position of two sides in two triangles is equal (such that if one side of one triangle is superimposed with one side of the other [triangle] or is parallel to it, then also the remaining sides of this one [triangle] come to lie on the remainder sides or parallel to it, thus the angles, which are lying between two such sides equal in their position, are equal" (*Ueber den Gebrauch*, p. 34–35).[230] Thus, Maimon's method of conversion can be used to detect whether the method of generalization can be applied to a given proposition.

I now turn to the question of whether Maimon's method is one of generalization or of universalization. Maimon uses the term *allgemein*, which can be translated as both *general* and *universal*. I believe his method is one of generalization and not of universalization since it offers abstraction from particular cases and not invention of a new form of proof or solution that can be applied to many particular cases. General [*allgemein*] things, according to Maimon, "arise through abstraction and the more we abstract, the more general the things become" (*Tr.*, p. 258). That is, in generalization, we abstract the general from the particular. For instance, the Pythagorean theorem: First, the Pythagorean theorem about a right-angled triangle is presented and proved and only then is the more general proposition regarding any triangle as such found. All Maimon's methods of invention are based on actual inventions and this method is no different: he does not prove a general case and from it prove the particular instances, but rather begins with particular propositions and then abstracts more general propositions from them. By using a method of universalization, by contrast, first the more general case is proven and from that it follows that all particular cases subsumed under it have been proven. Such is Leibniz's method of universalization wherein the object is to find a single analytic formula that can be applied to several different cases making the analysis or synthesis of each partic-

---

sides respectively, and have the angles contained by the equal straight lines equal, they will also have the base equal to the base, the triangle will be equal to the triangle, and the remaining angles will be equal to the remaining angles respectively, namely those which the equal sides subtend" (Heath 1956, Vol. I, p. 247).

230 Maimon's proposition is about two equal sides and not all the sides of two triangles, but it is similar to *Elements* VI.5: "If two triangles have their sides proportional, the triangles will be equiangular and will have those angles equal which the corresponding sides subtend" (Heath 1956, Vol. II, p. 202). In *Critical Investigations of the Human Spirit*, Maimon makes a similar claim, that the proposition about congruence between triangles is a corollary of the proposition regarding similarity between triangles (*KrU*, p. 362).

ular case redundant (Leibniz 1961a, p. 97). Leibniz presents an example: on a given point D, to construct a perpendicular DB on a given conic section ABC. The point D can be given in various places and the conic section can be varied as well (it can be a straight line, circular, parabolic, elliptic or hyperbolic). Each of these lines and all the possible places of point D require separate calculation. Leibniz adds that 35 different calculations are to be made, in order to get "one perfect solution of the proposed problem. And nevertheless, I claim to grasp them all in a sole calculation which will be no more difficult than the most difficult of the 35 cases" (Leibniz 1961a, p. 122–123). Leibniz's one universal formula offers solutions for all the particular cases that are subsumed under it. Maimon begins with a given particular proposition and only then searches for the more general proposition, whereas Leibniz's begins with the general rule and then seeks for its proof. For his own theory of invention, Maimon did not present a method of universalization, which is a method of analysis. Instead, he presents a method of generalization, which is a method of synthesis.[231] It is an outcome of his decision to extract methods of invention from actual inventions.

4.4 Assuming a Problematic Proposition as True

In a problematic proposition, the ground for the connection between the subject and predicate is not necessary and opaque to reason. Therefore, Maimon includes this method as part of his methods of synthesis. However, it can also be used as a method of analysis and is even presented as such by Maimon himself. He also refers to it as a method that can transform a problematic synthetic proposition into a necessary analytic one, thus making this method important. Unfortunately, his suggestion of how to conduct this transformation proves itself problematic.

In the articles published in 1795, Maimon refers to this method as one of analysis. He does not describe this method by itself as he does with the kinds of analysis or the other two mentioned methods of synthesis, conversion and generalization. Rather, he presents this method indirectly. When pointing out

---

[231] I am using here the definitions of analytic and synthetic knowledge as they appear in Maimon's third kind of analysis: in analytic knowledge, we begin with the universal and, since what is applied to the universal applies also to the particulars subsumed under it, once we prove the universal proposition we have proven the particular cases as well. In synthetic knowledge, we begin with proving all the particular cases and only by doing so do we arrive at the general proposition. The direction of proof in analysis is from the universal to the particular and in synthesis from the particular to the universal (*Ueber den Gebrauch*, p. 28–29).

that Schwab mistakenly considers analysis to be regressive only, Maimon describes the method as assuming the problematic proposition true: we assume proposition E to be a problematic conclusion assumed true prior to its proof and we deduce other premises from it until we arrive at "the self-evident or elsewhere proven proposition A, or to its opposite" (*Das Genie*, p. 377–378; *Ueber den Gebrauch*, p. 23). By this process, hypothesis E is demonstrated to be either true or false (*Das Genie*, p. 377–378; *Ueber den Gebrauch*, p. 23).[232] This method of assuming a problematic proposition as true and then, by using deduction, arrive at a known truth (or its contradiction) is used by Euclid in both *Elements* and *Data*. Proclus mentions that the difference between assuming a problematic proposition as true and an axiom is based on whether the proposition is demonstrated: When a proposition is assumed as true and is required to be demonstrated it is called an *assumption,* and when a proposition is assumed as true without demonstration it is called an *axiom* (Proclus 1789, p. 25). While Maimon does not make this distinction, we can assume that he would subscribe to its rationale. Thus, assigning this method to analysis is due to the demonstration it entails.

Maimon offers a criterion to verify whether propositions were given as problematic assumptions before being proven or whether they can be presented only after being proven and were not assumed beforehand. The need for such criterion is in the interest of inventing, rather than learning. According to him, although propositions are presented in mathematical textbooks before their proofs, in some cases the inventor's proof precedes the presentation of the proposition. Maimon's example of such a case in which a proof precedes the proposition (that is, that the proposition was not assumed as true before its proof) is *Elements* III.20. This proposition could not have been assumed before being proven since we had no ground on which to assume that the angle at the center of a

---

[232] In *Histoire des mathématiques*, which inspired Maimon's work, *A Short Exposition of Mathematical Inventions,* Montucla describes the method of assuming a proposition as true as a method of analysis in a manner that resembles Maimon's description: "La marche de l'Analyse est différente. Cette méthode est nécessaire, lorsqu'il s'agit de la recherche de quelque questions géométrique, soit problème, soit théorême. Ici l'on commence à prendre pour vrai ce qui est en question, ou l'on regarde comme résolu le problême qu'il s'en s'agit de resoudre. On tire de-là les conséquences qui s'en déduisent, et de celles-ci de nouvelles, jusqu'à ce que l'on soit parvenu à quelque chose de manifestement vrai ou faux, si c'est un théorême; de possible ou d'impossible à executer, si c'est un problême" (Montucla 2007, Vol. 1, p. 165). Whereas Montucla's description obtains for both theorems and problems, Maimon's description above addresses theorems alone. Only in *Methods of Invention* does Maimon refer to problems in relation to this method, and on that occasion he regards it as a method of synthesis. He briefly refers to the application of this method to problems and their resolution in the possibility or impossibility of what is assumed when he mentions *reductio ad absurdum* (*Erfindungsmethods*, p. 143–144).

circle is double the angle at the circumference when both angles have the same circumference at the basis (*Ueber den Gebrauch*, p. 20 – 21). Maimon also makes the observation that if a proposition is convertible, then it can precede its proof. The assumption of this problematic proposition as true originates already in the diagram.[233] Since the converse can serve as the given hypothesis as well, what was once the given becomes the sought. *Elements* I.5 is mentioned as an example of a proposition that is convertible and can therefore precede its proof (*Ueber den Gebrauch*, p. 20). Formulating this criterion, as to when one can use the method of assuming a problematic proposition as true, is an instantiation of Maimon's preference for invention over learning.

In *Method of Invention*, this method is presented as part of the methods of synthesis (*Erfindungsmethods*, p. 143–144). There, Maimon states that if we treat a problematic proposition as if it were true and by demonstration arrive at the identity of a thing with itself ($A = A$), then the problematic proposition is proven to be true (*Erfindungsmethods*, p. 144). This claim is problematic in a few aspects. But first, let us examine it: He begins his explanation by suggesting that in the conclusion, we arrive at an identity. Since the identity of a thing with itself is a necessary truth, the premise is also true. To a possible criticism, Maimon answers that he is aware that true conclusions can arise from false premises. However, he adds, a conclusion arising from a false premise is only a logical-formal truth and not a real consequence of the premise. The ground for the consequence is not the false premise, but some other ground. In the case where the conclusion is the identity of a thing with itself, it serves both as ground and real consequence, and therefore, the problematic proposition is proven true (*Erfindungsmethods*, p. 144–145). A few problems arise. First, in his use of identity he refers to a complete identity (identity of a thing with itself; $X = X$) and not equality ($A = B$). By that, he is assigning the necessity of a complete identity onto a mathematical problem.[234] However, proofs in Euclidean geometry do not resume in complete identities but in equalities. Maimon does not present an example. Since his work was not published, we can assume that he would have either presented an example or changed the argument. Due to Maimon's presentation of the method, both as a method of analysis and of synthesis, it is interesting to note that, just before claiming that an invented proposition can be reduced to complete identity, he assigns this method to synthesis. Most

---

233 Maimon does not use the term *diagram* but *perception* [*Augenschien*].
234 Here I follow Rabouin' distinction regarding Leibniz' "analysis of notions" into two different kinds of analysis. In logical analysis, notions are reduced to complete, or "absolute", identity (*A est A*). In mathematical analysis, notions are reduced to incomplete, or "hypothetical", identity (*Si A est B, et B est C, tunc A est C*; Rabouin 2013, p. 122; p. 130).

likely he does so since the invention of the problematic proposition is itself synthesis, even though the demonstration that follows it is analysis. Here Maimon does not extract a method from an actual invention but attempts to invent a new method by improving an existing one. This is but another example of Maimon's efforts to improve Euclidean geometry instead of working in another field of mathematics, such as algebra or analytical geometry. Considering Maimon's suggestion of arriving at a complete identity, this method would have been better applied to algebra since there we can arrive at the equation $X = X$ as a conclusion. The identity $X = X$ can serve as a sought in algebra, but propositions such as "angle ABC is equal to angle ABC" do not appear in Elements.[235] Also, equality of one thing to another is a common notion in Elements, but the identity of a thing with itself is not.[236] Most probably because there is no use of judgments expressing identity of a thing with itself. By suggesting that our demonstrations should end with complete identity, rather than equality, Maimon leaps from metaphysical truths to logical ones. Equality is between two determined objects that are different. A complete identity is the identity of a thing with itself; the object can be either determined or undetermined.[237] The transition from the metaphysical to the logical also has implications for the kind of necessity we attribute a proposition. Maimon suggests taking a problematic judgment and turning it into a necessary one. When he mentions necessity, he does not elaborate on whether this necessity is objective or subjective. Since the thought of any thinking being follows the principle of identity, complete identity is an objective truth.[238] However, in the case of an equality, such as the equality found in Elements I.5, the truth is only subjective (since it is grounded not only the principle of contradiction, but also on other grounds, such as intuition). Maimon's wish to assign objective necessity of the identity of a thing with itself to equality is problematic since he does not give an account of how this is justified.

---

235 When proving congruence between two triangles that share a common angle (as is the case in Elements I.5), in order to express that in the two compared triangles (triangle AFC and triangle AGB) there is an identical angle we do not show identity or equality, we use the notion of *being common* (or *sharing*): "[...] and they contain a common angle, the angle FAG" (Heath 1956, Vol. I, p. 251).
236 Elements, C.N.1: "Things which are equal to the same thing are also equal to one another" (Heath 1956, Vol. I, p. 155).
237 "If I think a and b as different, then I cannot think objects of thought in general under a and b, but only determinable objects, for an *objectum logicum* cannot differ from an *objectum logicum*, i.e. from itself" (Tr., p. 151).
238 "A truth recognized by any particular thinking being is to this extent merely a subjective truth. But it is an objective truth if this being recognizes it in such a way that every thinking being in general (in so far as it is a thinking being) must recognize it" (Tr., p. 151).

Maimon suggests that by using this method, what was proven apagogically (by the method of *reductio ad absurdum*) can be proven using direct syllogisms (*Erfindungsmethoden*, p. 144). This suggestion is not original but was already suggested by Aristotle.[239] Maimon does not give an example of how this can be done. However, in Section 4.1.6 I have presented an example of this exact case: Euclid proves Proposition 30 in *Data* by using *reductio ad absurdum* and this was later proven directly by Simson.

Lastly, Maimon suggests using this method when paraphrasing problems into theorems. If we change the problem "to find A which is B" into the hypothetical proposition "if A is B" and if we arrive at the consequence of the identity $X = X$, then we have proven that A must be B. He then concludes that instead of solving a problem, we have invented a theorem (*Erfindungsmethoden*, p. 145). Maimon's suggestion, to transform problems into propositions, would have been productive if in fact we could arrive at the sought identity, $X = X$ in Euclidean geometry. The idea of transforming problems into propositions is not original and there are in fact problems in *Elements* that are proven as propositions in *Data* (such as *Elements* I.22 and *Dt.* 39).[240] Were Maimon to give an example that proves his argument, it would have been more convincing. The problem in Maimon's claim rests in his appeal to complete identity and suffers from the obstacles described above.

The evolution of Maimon's method embodies several features of his theory of invention: first, attempting to prove a problematic proposition to be necessary using identity judgments can be seen not only as stemming from the choice to formulate a theory of invention as an art of finding arguments, but also as a reason for such a formulation (even if not expressed as such). Second, he considers this method at times analytic and at times synthetic, changing not so much the method itself but rather its focus. This reflects his general view of analysis and synthesis as complementary, with a shifting line of differentiation between the

---

[239] "Syllogisms by *reductio ad impossibile* are the same <in this respect> as direct syllogisms. For they result through the consequents and antecedents of each term. And in both cases the investigation is the same, since if something is proved directly it can also be proven with a syllogism through impossibility through the same terms, and what is proved through impossibility can also be proved directly" (Aristotle, *Prior Analytics*, 45a23–8, in: Alexander of Aphrodisias, 2006, p. 107). Aristotle also adds that the distinction between direct syllogism and proof by *reductio ad absurdum* lies in whether one or two of the premises are true: "A direct syllogism is different from an *ad impossibile* one because in a direct one both premises are posited truly, [but in an *ad impossibile* one is posited falsely]" (Aristotle, Prior Analytics, 45b8–11, in: Alexander of Aphrodisias, 2006, p. 112).

[240] The problem presented in *Elements* I.22 and its corresponding proposition, *Dt.* 39 were discussed in Section 4.1.2.

two. Hence, we find a method of construction which he names *analysis* (analysis of the object) on the one hand, and a method of synthesis that ends with identity judgment (assuming a problematic proposition as true) on the other hand. The latter gives us a clue as to what Maimon eventually considered invention to be: logical analysis. This in spite of the fact that, as a method, logical analysis has been proven to be problematically applied on examples, and even though Maimon states that logic is insufficient for the task of invention. His theory of invention is meant to reveal connections between truths that cannot be revealed using the tools general logic has to offer. But, as the last method shows, this does not mean that these connections themselves are not logical.

Conclusion

In answer to the question, "What is invention?" Maimon has many responses: finding unknown connections between truths, forming syllogisms and finding a new object or attribute of an object. Invention is also constructing a new object in intuition (e.g. the caustic curve) or in experience (e.g. Von Guericke's air pump). Invention is described as the action of the faculties of invention, reason, understanding and imagination. This complexity is probably the reason behind Maimon's tendency to define concepts such as *invention*, *given*, *analysis* and *synthesis* in both narrow and broad senses, even if at times he does not give an account for doing so. Tension between his appeal to logic on the one hand, and to sensibility on the other hand, is present in both his theory and methods. On the one hand, his theory of invention is an art of forming syllogisms and the purpose of his methods is to find premises, conclusions or middle terms for these syllogisms. On the other hand, his theory together with its methods is based on actual inventions, leaving out other methods of invention that are based on symbolic cognition alone. The foundations of his theory are set in logic, but since that is insufficient for a theory of invention, he turns to sensibility for the purpose of expanding knowledge.

The idea that invention consists of forming syllogisms appears in Maimon's work on the differences between the genius and the methodical inventor. While both types of inventors search for propositions to serve as premises or conclusions, the methodical inventor can account for each step taken whereas the genius cannot. For this reason, Maimon's theory of invention is designed to improve the work of the methodical inventor rather than that of the genius. Even though many philosophers of his time elevated the status of the genius, Maimon prioritized the methodical inventor, whose character had been celebrated in the prior century. Some characteristics of Maimon's conception of the genius and the methodical inventor can be traced back to Kant, Gerard and Batteux on the one hand and to Bacon, Descartes and Leibniz on the other. However, Maimon's viewpoint was like no other. The most notable part of Maimon's conception is that the genius is a scientific genius who forms syllogisms (even if unconsciously) and is distinguished from the methodical inventor only subjectively. The re-elevation of the status of the methodical inventor and his stance that the genius is a scientific genius could have resulted from his initial stand that to invent is to form syllogisms.

Maimon's theory of invention is meant to offer means of invention that general logic and the art of syllogism cannot. Nevertheless, his notion of invention in general is one of conceiving and forming syllogisms, notwithstanding that this theory together with its methods is based on invented objects given in intuition. Although not explicitly mentioned as such, it is evident that Maimon's *ars inveniendi* is an art of finding arguments. While thoroughly acquainted with early modern theories of invention (such as Bacon's, Descartes' and Leibniz's) and with more contemporary works (such as Lambert's), it seems that Maimon's greatest influence is Aristotle and the art of finding arguments. Accordingly, Maimon chose Euclidean geometry to serve as the exemplar science of his theory of invention rather than analytic geometry, *ars characteristica* or even calculus. I offer two possible reasons for this preference: First, by using an art of finding propositions to serve as premises or conclusions we may arrive at identity judgments such as those found in the last method (Section 4.4). Based on the necessity of the identity judgment at the end of the demonstration, Maimon claims that a problematic proposition assumed as true is proven necessary. Thus, defining the action of inventing as the formulation of syllogisms may be based on a desire to show that mathematical truths can be reduced to logic. Secondly, I suggest that in choosing Euclidean geometry over more modern fields of mathematics, Maimon wished to avoid the use of symbolic cognition and of objects that may be real, but not actual. This avoids the question of whether the invented objects are possible since what is actual is ipso facto possible. Euclidean geometry, based on objects given in actuality and including propositions whose ground is intuition, is therefore the best mathematical field for his methods of invention. For this reason, the notion of geometry that arises from Maimon's work on invention is of a science of figures and not one of space, despite the fact that the latter notion of geometry is the one arising from his philosophy in general.

Upon examining Maimon's methods of invention I find that many are similar to the problem-solving practices of Greek geometricians, with most being mentioned in Proclus' commentary to Book I of *Elements*. Nevertheless, Maimon has improved upon these methods and at times offered an articulation of a method where only the practice was found, as is the case with his fourth kind of analysis. In following Maimon's methods and examples with my own examples, I have discovered new aspects of his methods, or in the case of his method of logical analysis, even a redefinition of the method. In the third method, for instance, I suggest an analysis of a number of cases and a method of amendment of a definition in order to present a new case of a problem or proposition. Following Maimon's work on conversion, I suggest a new criterion for differentiating between logical and geometrical conversions: in logical conversion, we interchange the subject with the predicate whereas in geometrical conversions the subject remains the same and only its predicates change.

A case study of the tension between the actions of the faculty of imagination on one side, and the actions of the faculties of understanding and reason on the other, is the method of analysis of the object. The essence of this method is construction, but this construction is meant to aid in the formulation of a syllogism – a syllogism wherein the given object is transformed into the object of the conclusion.

The relations between logic and analysis in its broader sense are illustrated in Maimon's method of logical analysis: on the one hand, this method is declared as necessary to any solution or proof, thus being highly important to invention. On the other hand, it is the other six kinds of analysis that enable finding propositions that serve as premises or conclusions in demonstrations. As for the application of the method itself, I argue that even though there are two methods found under the title of *logical analysis* – the method of regressive analysis and the method of forming syllogism – neither can claim to be necessary in any proof or solution. In my application of these two methods on the same proposition (*Dt.* 1), some problems arise. Therefore, we should either invalidate Maimon's claim that logical analysis is indispensable or, better yet, consider that what is meant by the term *logical analysis* is in fact the principle of contradiction.

The importance of using mathematical examples when presenting mathematical methods is highlighted by the last method. As it is one thing to claim that problematic truths can be turned into necessary ones by arriving at identity judgments in demonstrations, but quite another to present concrete examples that prove propositions to be objectively necessary. Maimon's wish to bridge Euclidean geometry and logic may have caused him to ignore the fact that an identity of a thing with itself is not found in either *Elements* or *Data*. Only equality of magnitudes is found.

This work includes many definitions of terms in both their narrow and broader sense, among them *invention, analysis, synthesis, given* and *a priori*. Most are presented as such by Maimon, some are my own. This duality results from the need to reconcile conceptual work, which attempts to be as clear and concise as possible, with practical work being done with these concepts. So, for instance, Maimon's use of the concept of *given* includes a meaning broader than the one he explicitly defines. In most cases, the dual meanings of a term refer to its purely conceptual notion and a notion based on intuition. This is a practical expression of Maimon's approach to the relation between understanding and sensibility as being continuous, not dichotomous:

Given: I suggest that the term *given* in the broader sense should be defined as "something which is presented to our cognition passively." This general term includes the terms *known* and *given* in their narrow senses. *Known* in the narrow sense means "given as a proposition". *Given* in the narrow sense means "given in intuition" and is the common meaning of the term. Even though Maimon does not make this differentiation, the need for the broader sense of *given* and the narrow notion of *known* arises from his use of the term *given* in the context of knowledge that we do not actively think. Furthermore, Maimon's use of *given* in intuition could have benefited from adopting the division of *given* as it appears in Euclid's *Data* (most notably, given in magnitude, in position and in form).

Invention: Invention in the narrow sense is defined by Maimon as presenting a new object *a priori*, whereas ascribing a new attribute to a given object *a priori* is *discovery*. Invention in the broader sense is forming syllogisms (either as an analytic or synthetic process) and includes both invention (in the narrow sense) and discovery. In Maimon's philosophy, invention has an even broader sense – "the introduction of something new": It could be the introduction of an object in experience, such as the telescope or the discovery of a new planet, or it could be the invention of a new mathematical object which is only symbolic cognition, such as the pure unit. This third notion of invention appears in Maimon's work, but not in the texts explicitly devoted to his theory of invention. Invention is also defined by Maimon as different kinds of logical analysis and synthesis (to which I refer below).

Analysis: Analysis in the narrow sense is defined as educing a predicate from the concept of the subject. It is grounded in the principle of contradiction alone. Analysis in the broader sense has two meanings and applications: one mathematical and the other philosophical. In both, analysis is not grounded in the principle of contradiction alone. In mathematical practice, this notion can be derived from the first six methods of invention which Maimon calls "kinds of analysis." These methods resemble *diorism*, a method of analysis used by Greek geometers. It includes practices such as finding all the cases of a given problem or verifying whether a condition is necessary. In philosophy, I also refer to analysis in the broader sense as *ampliative analysis*. It refers to the notion of analysis presented in his *Logic*: educing a predicate not from the concept of the subject but rather from the object. This kind of analysis is also grounded in intuition. Analysis in general is manifold in unity.

Logical analysis: Maimon presents two kinds of logical analysis: the first is as forming syllogisms, grounded in the principle of contradiction alone. It is based on categorical propositions. The second kind of logical analysis is also called *mathematical analysis* and is grounded both in intuition and the principle of contradiction. It is based on hypothetical propositions (that refer to determined objects). Determining whether categorical propositions can be reduced to hypothetical propositions, or vice versa, can indicate which of the two notions of logical analysis is more fundamental in Maimon's view. Although Maimon mentions both directions of transformation of these two kinds of judgments, in essence he regards all propositions as categorical. Even though in his work he often refers to invention as analysis in the broader sense, his fundamental notion of invention is that of logical analysis based on the principle of contradiction alone.

Analysis of the object: There are three kinds of analysis of the object, all of which involve the presentation of the object in intuition: (I) analysis of the object as educing a predicate immediately from the object (the understanding produces judgments); (II) analysis of the object as educing a new predicate indirectly by using demonstration (reason produces inferences); (III) analysis of the object as the method of transformation of the given object (the faculty of imagination produces a new construction based on a given construction of a given object). In cases of the first kind (such as in the judgment "a triangle has three angles"), it can be assumed that although Maimon assigned necessity to this kind of judgment, he meant to assign only subjective, not objective, necessity. Immediate educement of the predicate can be explained by the type of determination between the subject and predicate whereas indirect educement of the predicate (as in the Pythagorean Theorem) can be explained by using the form of conclusions of hypothetical judgments. In this case, the relation of determinability between the subject and predicate is not immediate but indirect.

Synthesis: Synthesis in the narrow sense is the introduction of a new object *a priori*. Synthesis in the broader sense is not defined, but implicitly includes a range of cases. In relation to his work on invention, synthesis is defined as beginning from something given: it can be a determinable, an object given in intuition or a proposition wherein the ground of the connection between the subject and predicate is not in the concept. This broader notion of synthesis also includes the introduction of a new predicate (i.e., what Maimon at times calls *analysis of the object*). Synthesis in general is unity in manifold.

Maimon does not mention his principle of determinability in his works on invention. This principle, verifying whether a synthesis of a subject and a predicate is real thought, has three criteria: being an absolute concept, having an object given in intuition and producing new consequences. Following my example of *right-angled trapezium*, I suggest omitting Maimon's third criterion and maintaining only the first two. The second criterion can be changed into presenting a *new* object in intuition.

A priori: Absolutely *a priori* is a cognition that precedes the cognition of the object and is grounded in the principle of contradiction alone. Therefore, according to Maimon, when the cognition of the objects is prior to cognition of the relation between the objects, it is *a posteriori* (formally, not materially). *A priori* in the broader sense is a cognition that precedes any sensation. These definitions of *a priori* have already been discussed by Maimon commentators.

The introduction of new definitions of synthesis (in the narrow sense) and analysis (in the broader sense) led Maimon to present a new version of Kant's question concerning the possibility of synthetic *a priori* judgments: "How are synthetic, or analytic judgments, whose predicate is not thought as contained in the concept of the subject, but rather in the subject itself, possible?" This question centers around the possibility of knowledge that is not absolutely *a priori* and not pure (in Maimon's terms, *pure* is only a product of the understanding). Kant's synthetic *a priori* knowledge is divided into two types of knowledge: synthetic and ampliative analytic. In synthetic judgments (e.g. "a triangle can be right-angled"), a new object arises and the relation of a determinant to the determinable is only possible and known not to be necessary. This kind of judgments cannot be transformed into apodictic judgments (wherein the predicate is necessarily contained in the subject) without a change of quantifier. That is, synthetic judgments in the narrow sense cannot be shown to be analytic in the narrow sense. In ampliative analytic judgments (e.g. "a triangle has three angles"), a new property is found (where the connection between subject and predicate is found in the object) and the relation of the determinant to the determinable is necessary. At times Maimon considers this kind of judgments as apodictic and at others as assertoric, but in either case it is not problematic. Therefore, it is not impossible to show that the predicate is contained in the concept of the subject and not only connected to its object. Hence, the application of Maimon's definitions of synthesis in the narrow sense and analysis in the broader sense leads to the conclusion that the first cannot be shown to be analytic in the narrow sense whereas the latter can.

Comparison of Maimon's definitions of the two pairs of concepts: analysis and discovery (both defined as "introducing a new attribute") and synthesis and invention (both defined as "introducing a new object") raises the question of whether the difference between discovery and invention is indeed only subjective as Maimon asserts and, therefore, is also the difference between analysis and synthesis. Were the difference real, Maimon should have presented for each a separate theory and method, of invention and of discovery. He also should not have interchanged the two terms, as he often does. Supporting the claim that this difference is only subjective are methods discussed in Chapter 4, such as the second kind of analysis, used for both invention and discovery. If we assume this difference to be subjective, it is inferred from the comparison of definitions that the difference between analysis (in the broader sense) and synthesis (in the narrow sense) is subjective as well. Since the difference between synthesis and analysis is based on different relations of determination – possible and not necessary in the first and necessary or yet unknown in the second – then using these definitions our finite understanding cannot overcome this difference.

The solution I offer is to turn to the idea of infinite understanding as creating real objects by thinking their concepts, thinking all possible determinations, together with an appeal to the notion that it is only in degree that our finite understanding differs from the infinite understanding. Being created in the image of God means that the finite understanding creates new concepts and objects and educes new attributes and properties from them. But, unlike the infinite understanding, it does so as different processes, and in the forms of intuition, space and time. Our finite understanding is either given the determinable and thus seeks for its determinant or is given the determination and seeks the determinant. What for the finite understanding is given and sought, is for the infinite understanding only thought.

To shed light on connections between truths, our finite understanding needs to be assisted by methods – some conceptual and some conducted in the forms of intuition. Maimon views these connections as essentially logical, even if in many cases we can only cognize them using intuition and not logical principles. Maimon's contribution to the process of human perfectibility and the advancement of knowledge may be humble but it is not insignificant. His ways of analysis of conceptual systems and mathematical objects are pertinent not only to Euclidean geometry or 18th century philosophy, but also to problem-solving in other fields of knowledge. First and foremost, Maimon's theory of invention offers thought-sharpening tools applicable even to the present-day.

Bibliography

Acerbi, Fabio (2011): "The Language of the 'Givens': Its Forms and Its Use as a Deductive Tool in Greek Mathematics". In: *Archive for the History of Exact Science* 65, pp. 119–153.
Adelung, Johann Christoph (1811): *Grammatisch-kritisches Wörterbuch der hochdeutschen Mundart*. Wien: Bauer.
Alexander of Aphrodisias (1991): *On Aristotle's Prior Analytics 1.1–7*, Tr. Barnes Jonathan. Ithaca, N.Y.: Cornell University Press.
Alexander of Aphrodisias (2001): *On Aristotle's Topics 1*. Van Ophuijsen, Johannes M. (Tr.). Ithaca, N.Y.: Cornell University Press.
Alexander of Aphrodisias (2006): *On Aristotle's Prior Analytics 1.23–31*. Mueller Ian (Tr.). Ithaca, N.Y.: Cornell University Press.
Andersen, Kirsti (1985): "Cavalieri's Method of Indivisibles". In: *Archive for History of Exact Sciences* 31. No. 4, pp. 291–367.
Aristotle (1976): *Posterior Analytics, Topica*. Tredennick, Hugh/Forster, Eduard Seymour (Trs.). Cambridge, Mass.: Harvard University Press.
Arnauld, Antoine/Nicole, Pierre (1996): *Logic or the Art of Thinking: Containing, besides Common Rules, Several New Observations Appropriate for Forming Judgment*. Jill Vance Buroker (Tr. & Ed.). Cambridge: Cambridge University Press.
Atlas, Samuel (1964): *From Critical to Speculative Idealism: The Philosophy of Solomon Maimon*. The Hague: Martinus Nijhoff.
Atlas, Samuel (1969): "Solomon Maimon's Doctrine of Fiction and Imagination". In: *Hebrew Union College Annual* 40/41, pp. 363–389.
Bacon, Francis (1859): "Great Instauration". In: *The Works of Francis Bacon, with a Life of the Author*. Vol. 3. Basil Montagu (Ed.). Esquire, Parry & McMillan: Philadelphia, pp. 329–522.
Batteux, Charles (1746): *Les beaux arts reduits a un même principe*. Paris: Durand.
Beeley, Philip (2008): "Infinity, Infinitesimals, and the Reform of Cavalieri: John Wallis and His Critics". In: Goldenbaum Usrula/Jesseph Douglas (Eds.): *Infinitesimal Differences: Controversies Between Leibniz and His Contemporaries*. Berlin, New York: Walter de Gruyter, pp. 31–52.
Ben Makhir, Jacob (1272): *Euclid's Data* [ספר המתנות לאקלידס]. The National Library of Israel. Comunita Israelitica di Mantuva, Ms. ebr. 3. Copied in 1612 [in Hebrew].
Bergman, Samuel Hugo (1967): *The Philosophy of Solomon Maimon*. Noah J. Jacobs (Tr.). Jerusalem: Magnes Press.
Bernard, Alain (2003a): "Sophistic Aspects of Pappus's *Collection*". In: *Archive for History of Exact Sciences* 57. No. 2, pp. 93–150.
Bernard, Alain (2003b): "Ancient Rhetoric and Greek Mathematics: A Response to a Modern Historiographical Dilemma". In: *Science in Context* 16. No. 3, pp. 391–412.
Bernard, Alain (2010): "L'arrière-plan rhétorique de la théorie de l'activité mathématique chez Proclus". In: Alain Lernoult (Ed.): *Études sur le commentaire de Proclus au 1er livre des Éléments d'Euclide*. Lille: Presses universitaires du Septentrion, pp. 67–85.
Boehm, Omri (2013): "Enlightenment, Prophecy, and Genius". In: *Graduate Faculty Philosophy Journal* 34. No. 1, pp. 149–178.

Bogomolny, Alexander (2016): "Pythagorean Theorem and its many proofs", in: *Interactive Mathematics Miscellany and Puzzles*. http://www.cut-the-knot.org/pythagoras/index.shtml, Visited on 24 May 2016.

Bos, Henk, J.M. (1974): "Differentials, Higher-Order Differentials and the Derivative in the Leibnizian Calculus". In: *Archive for History of Exact Sciences* 14. No. 1, pp. 1–90.

Bransen, Jan (1991): *The Antinomy of thought: Maimonian Skepticism and the Relation between Thoughts and Objects*. Dordrecht: Kluwner Academic Publishers.

Buchenau, Stefanie (2013): *The Founding of Aesthetics in the German Enlightenment: The Art of Invention and the Invention of Art*. Cambridge; New York: Cambridge University Press.

Buzaglo, Meir (2002): *Salomon Maimon: Monism, Skepticism and Mathematics*. Pittsburgh, PA: University of Pittsburgh Press.

Capozzi, Mirella (2006): "Kant on Heuristics as a Desirable Addition to Logic". In: Carlo Cellucci/Paolo Pecere (Eds.): *Demonstrative and Non-Demonstrative Reasoning in Mathematics and Natural Science*. Cassino: Edizioni dell'Università degli Studi di Cassino, pp. 123–181.

Cifoletti, Giovanna (2006): "Mathematics and Rhetoric. Introduction". In: *Early Science and Medicine* 11. No. 4, pp. 369–389.

Claessens, Guy (2009): "Clavius, Proclus, and the Limits of Interpretation: Snapshot-Idealization *versus* Projectionism". In: *History of Science* 47. No. 3, pp. 317–336.

Corr, Charles A. (1972): "Christian Wolff's Treatment of Scientific Discovery". In: *Journal of the History of Philosophy* 10. No. 3, pp. 323–334.

De Risi, Vincenzo (2007): *Geometry and Monadology: Leibniz's Analysis Situs and Philosophy of Space*. Basel: Birkhäuser.

De Risi, Vincenzo (Ed.) (2015): *Mathematizing Space: The Objects of Geometry from Antiquity to the Early Modern Age*. Basel: Birkhäuser.

Descartes, René (1925): *The Geometry of René Descartes* (1st ed. 1637). David Eugene Smith/Marcia L. Latham (Tr.). Chicago, London: The Open Court Publishing Company.

Descartes, René (1985): *The Philosophical Writings of Descartes*, Vol. 1. John Cottingham/Robert Stoothoff/Dugald Murdoch. (Trs.). Cambridge: Cambridge University Press.

Diderot, Denis/d'Alembert, Jean le Rond (Eds.) (2016): *Encyclopédie, ou dictionnaire raisonné des sciences, des arts et des métiers, etc*. Robert, Morrissey and Glenn, Roe (Eds.) University of Chicago: ARTFL Encyclopédie Project (Spring 2016 Edition). http://encyclopedie.uchicago.edu/.

Fichte, Johann Gottlieb (1988): "Concerning the Concept of the Wissenschaftslehre (1st ed. 1794)". In: *Fichte: Early Philosophical Writings*. Daniel Breazeale (Tr.). New York: Cornell University Press, pp. 94–135.

Flögels, Carl Friedrich (1760): *Einleitung in die Erfindungskunst*. Breslau and Leipzig: Johann Ernst Meyer.

Fournarakis, Philippos/Christianidis, Jean (2006): "Greek Geometrical Analysis: A New Interpretation Through the "Givens"-Terminology". In: *Bolletttino Di Storia delle Scienze Matematiche* 26. No. 1, pp. 33–56.

Franks, Paul (2003): "What Should Kantians Learn From Maimon's Skepticism?". In: Gideon Freudenthal (Ed.): *Salomon Maimon: Rational Dogmatist, Empirical Skeptic: Critical Assessments*. Dordrecht; Boston: Kluwer Academic, pp. 200–232.

Freudenthal, Gideon (2004): "Salomon Maimon: Philosophizing in Commentaries". In: *Daat: A Journal of Jewish Philosophy & Kabbalah* 53. pp. 125–160 [in Hebrew].

Freudenthal, Gideon (2006): *Definition and Construction: Salomon Maimon's Philosophy of Geometry*. Preprint 317 of the Max Planck Institute for the History of Science, Berlin.
Freudenthal, Gideon (2010): "Maimon's Philosophical Program, Understanding versus Intuition". In: Fred Rush/Jürgen Stolzenberg (Eds.): *International Yearbook of German Idealism* 8: *Philosophy and Science*. Berlin: de Gruyter, pp. 83–105.
Freudenthal, Gideon (2011): "Salomon Maimon's Development from *Kabbalah* to Philosophical Rationalism". In: *Tarbiz* 80. No. 1, pp. 105–71 [in Hebrew].
Freudenthal, Gideon/Klein-Braslavy Sara (2003): "Salomon Maimon reads Moses Ben-Maimon: On Ambiguous Names". In: *Tarbiz* 72. No. 4, pp. 581–613 [in Hebrew].
Gardies, Jean-Louis (2001): *Qu'est-ce que et pourquoi l'analyse? Essay de definition*. Paris: J. Vrin.
Gerard, Alexander D.D. (1774): *An Essay on Genius*. London, Edinburgh: W. Strahan; T. Cadell; W. Creech.
Goethe, Norma B/Beeley Philip/Rabouin David (Eds.) (2015): *G. W. Leibniz, Interrelations Between Mathematics and Philosophy*. Dordrecht: Springer.
Green-Pedersen/Niels Jørgen (1984): *The Tradition of the Topics in the Middle Ages: The Commentaries on Aristotle's and Boethius' Topics*. Munich and Vienna: Philosophia.
Green-Pedersen/Niels Jørgen (1987): "The Topics in Medieval Logic". In: *Argumentation* 1. No. 4, pp. 407–417.
Grosholz, Emily R. (2005): *Cartesian Method and the Problem of Reduction*. Oxford, New York: Oxford University Press.
Gueroult, Martial (1929): *La Philosophy Transcendantale de Salomon Maïmon*. Paris: F. Alcan.
Guyer, Paul (2011): "Gerard and Kant: Influence and Opposition". In: *The Journal of Scottish Philosophy* 9. No. 1, pp. 59–93.
Harari, Orna (2004). *Knowledge and Demonstration: Aristotle's Posterior Analytics*. Dordrecht, Boston, London: Kluwer.
Heath, Thomas Little (1956): *The Thirteen Books of Euclid's Elements* (1st ed. 1908). New York: Dover Pub, 2nd ed.
Herrera, Hugo E. (2010): "Salomon Maimon's Commentary on the Subject of the Given in Immanuel Kant's Critique of Pure Reason". In: *The Review of Metaphysics* 63. No. 3, pp. 593–613.
Hintikka, Jaakko (1992): "Kant on the Mathematical Method". In: Carl J. Posy (Ed.): *Kant's Philosophy of Mathematics: Modern Essays*. Dordrecht; Boston: Kluwer Academic, pp. 21–42.
Hintikka, Jaakko (1998): "What is Abduction? The Fundamental Problem of Contemporary Epistemology". In: *Transactions of the Charles S. Pierce Society* 34. No. 3, pp. 503–533.
Hintikka, Jaakko (2012): "Method of Analysis: A Paradigm of Mathematical Reasoning?". In: *History and Philosophy of Logic* 33. No. 1, pp. 46–67.
Hintikka, Jaakko/Remes, Unto (1974): *The Method of Analysis: Its Geometrical Origin: Its General Significance*. Dordrecht: D. Reidel.
Hoyningen-Huene, Paul (2008): "Systematicity: the nature of science". In: *Philosophia* 36. No. 2, pp. 167–180.
Illusie, Luc (2004): "What is… a Topos?". In: *Notices of the American Mathematical Society* 5. No. 9, pp. 1060–1061.
Jaffe, Kineret S. (1980): "The Concept of Genius: Its Changing Role in Eighteenth-Century French Aesthetics". In: *Journal of the History of Ideas* 41. No. 4, pp. 579–599.

Jardine, Lisa (1974): *Francis Bacon: Discovery and the Art of Discourse*. London: Cambridge University Press.
Jefferson, Ann (2009): "Genius and its Others". In: *Paragraph* 32. No. 2, pp. 182–196.
Kant, Immanuel (1983): *Critique of Pure Reason*. (1st ed. 1780, 1787). Kemp Smith, Norman (Tr.) London: Macmillan, (2nd ed.).
Kant, Immanuel (1987): *Critique of Judgment*. (1st ed. 1790). Werner S. Pluhar (Tr.) Indianapolis/Cambridge: Hackett Publishing Company.
Kant, Immanuel (1963): *Kant's Introduction to Logic, and his Essay on the Mistaken Subtilty of the Four Figures*. (1st ed. 1800). Thomas Kingsmill Abbott (Tr.), London: Vision.
Kästner Abraham Gotthelf (1758): *Anfangsgründe der Arithmetik, Geometrie, ebenen und sphärischen Trigonometrie, und Perspectiv. Der mathematischen Anfangsgründen ersten Theils erste Abtheilung*. Göttingen, Vandenhoek.
Kästner Abraham Gotthelf (1781): *Anfangsgründe der angewandten Mathematik: Zweither Theil, Zweyte Abtheilung: Astronomie, Geographie, Chronologie und Gnomonik*. Göttingen, Vandenhoek.
Kästner Abraham Gotthelf (1786): *Anfangsgründe der Mathematik, Isten Theils erste Abteil: Anfangsgründe der Arithmetik, Geometrie ebenen und sphärischenen Trigonometrie und Perspectiv*. Göttingen: Vandenhoeck.
Katzoff, Charlotte (1981): "Salomon Maimon's Critique of Kant's Theory of Consciousness". In: *Zeitschrift für philosophische Forschung* 35. No. 2, pp. 185–195.
Kienpointer, Manfred (1997): "On the Art of Finding Arguments: What Ancient and Modern Masters of Invention Have to Tell Us About the *Ars Inveniendi*". In: *Argumentation* 11. No. 2, pp. 225–236.
Kluge, Friedrich (1960): *Etymologisches Wörterbuch der Deutschen Sprache* (1st ed. 1894). Strassburg: K.J. Trubner, (18th ed.).
Knobloch, Eberhard (2010): "Leibniz Between *ars characteristica* and *ars inveniendi*: Unknown News About Cajori's 'Master-Builder' of Mathematical Notations." In: Albrecht Heeffer/Maarten Van Dyck (Eds.): *Philosophical Aspects of Symbolic Reasoning in Early Modern Mathematics*. London: College Publications, pp. 289–302.
Knorr, Wilbur Richard (1993): *The Ancient Tradition of Geometric Problems* (1st ed. 1945). New York: Dover Publications, (2nd ed.)
Kuntze, Friedrich (1912): *Die Philosophie Salomon Maimons*. Heidelberg: C. Winter.
Lachterman, David (1992): "Mathematical Construction, Symbolic Cognition and the Infinite Intellect: Reflections on Maimon and Maimonides". In: *Journal of the History of Philosophy* 30. No. 4, pp. 497–522.
Lambert, Johann Heinrich (1764): *Neues Organon oder Gedanken über die Erfoschung und Bezeichnung des Wahren und dessen Unterscheidung vom Irrthum und Schein. Erster Band*. Leipzig: Johann Wendler.
Larsen, Elizabeth (1993): "Re-Inventing Invention: Alexander Gerard and *An Essay on Genius*". In: *Rhetorica: A Journal of the History of Rhetoric* 11. No. 2, pp. 181–197.
Leff, Michael (1996): "Commonplaces and Argumentation in Cicero and Quintilian". In: *Argumentation* 10. No. 4, pp. 445–452.
Leibniz, Gottfried Wilhelm (1765a): "Discours touchant la methode de la certitude et l'art d'inventer (1st ed. 1690)". In: *Oeuvres philosophiques latines & françoises de feu Mr. de Leibnitz: tirées de ses manuscrits qui se conservent dans la Bibliothèque Royale à Hanovre, et publiées par Mr. Rud. Eric Raspe, avec une préface de Mr. Kaestner,*

Professeur en mathématiques à Göttingen. Rudolph Erich Raspe (Ed.). Amsterdam, Leipzig: Jean Schreuder, pp. 523–532.

Leibniz, Gottfried Wilhelm (1765b): "Nouveaux essaies sur l'entendement humain (1704)". In: *Oeuvres philosophiques latines & françoises de feu Mr. de Leibnitz: tirées de ses manuscrits qui se conservent dans la Bibliothèque Royale à Hanovre, et publiées par Mr. Rud. Eric Raspe, avec une préface de Mr. Kaestner, Professeur en mathématiques à Göttingen*. Rudolph Erich Raspe (Ed.). Amsterdam, Leipzig: Jean Schreuder, pp. 1–496.

Leibniz, Gottfried Wilhelm (1961a): "De la méthode de l'universalité (1st ed. 1674)". In: *Opuscules et fragments inedits de Leibniz: extraits des manuscrits de la Bibliothèque Royale de Hanovre*. Louis Couturat (Ed.). Hildesheim: G. Olms, pp. 97–143.

Leibniz, Gottfried Wilhelm (1961b): "Projet d'un art d'inventer: projet et essais pour arriver à quelque certitude pour finir une bonne partie des disputes et pour avancer l'art d'inventer (1st ed. 1686)". In: *Opuscules et fragments inedits de Leibniz: extraits des manuscrits de la Bibliotheque Royale de Hanovre*. Louis Couturat (Ed.). Hildesheim: G. Olms, pp. 175–182.

Leibniz, Gottfried Wilhelm (1961c): "Nouvelles ouvertures (1st ed. 1686)". In: *Opuscules et fragments inedits de Leibniz: extraits des manuscrits de la Bibliothèque Royale de Hanovre*. Louis Couturat (Ed.). Hildesheim: G. Olms, pp. 224–229.

Leibniz, Gottfried Wilhelm (1969a): "Two Studies in the Logical Calculus (1st ed. 1679)". In: *Philosophical Papers and Letters*. Leroy E. Loemker (Ed. & Tr.). Dordrecht: D. Reidel. (2nd ed.), pp. 235–239.

Leibniz, Gottfried Wilhelm (1969b): "Studies in a Geometry of Situation with a Letter to Christian Huygens (1st ed. 1679)". In: *Philosophical Papers and Letters*. Leroy E. Loemker (Ed. & Tr.). Dordrecht: D. Reidel. (2nd ed.), pp. 248–258.

Leibniz, Gottfried Wilhelm (1983): "Nouvelle méthode pour les maxima et les minima, et de même pour les tangentes, qui ne s'oppose ni aux quantités fractionnaires, ni irrationnelles, et un genre de calcul pour eux (1st ed. 1684)". *Acta Eruditorum*, October 1684. In: *Oeuvre concernant le calcul infinitesimal: suivi du recueil de diverses pièces sur la dispute entre Leibnitz et Newton d'après Desmaiseaux, et de fragments du traité des sinus du quart de Cercle de Pascal; de la méthode du maximum et du minimum de Fermat*. Jean Peyroux (Tr.). Paris: Blanchard, pp. 4–9.

Leibniz, Gottfried Wilhelm (1998): *Recherches générales sur l'analyse des notions et des vérités: 24 thèses métaphysiques et autres textes logiques et métaphysiques: Introd. and notes: Jean-Baptiste Rauzy*. Cattin, Emmanuel/Clauzade, Laurent/De Buzon, Frédéric (Trs.). Paris: Presses Universitaires de France.

Lévy, Tony (1997): "The Establishment of the Mathematical Bookshelf of the Medieval Hebrew Scholar: Translations and Translators". In: *Science in Context* 10. No. 3, pp. 431–451.

Lorenz, Johann Friedrich (1824): *Euklid's Elemente, funfzehn Bücher* (1st ed. 1781). Halle: Waisenhauses.

Mach, Ernst (1896): "On the Part Played by Accident in Invention and Discovery". In: *The Monist* 6. No. 2, pp. 161–175.

Mahoney, Michael S. (1968): "Another Look at Greek Geometrical Analysis". In: *Archive for History of Exact Sciences* 5. No. 3–4, pp. 318–348.

Maimon, Salomon (1778): *Hesheq Shelomo*. The National Library of Israel, F45617, Ms Heb. 8°6426 [in Hebrew].

Maimon, Salomon (1791): *Philosophisches Wörterbuch, oder Beleuchtung der Wichtigsten Gegenstände der Philosophie, in alphabetischer Ordnung. Erste Stück*. Berlin: Johann Friedrich Unger.
Maimon, Salomon (1793): *Bacons von Verulam Neues Organon. Aus dem Lateinischen übersetzt von George Wihelm Bartoldy. Mit Anmerkungen von Salomon Maimon. Zwei Bände. Mit Kupfern*. Berlin: Gottfried Carl Nauclk.
Maimon, Salomon (1794): *Die Kathegorien des Aristoteles, mit Anmerkungen erläutert und als Propädeutik zu einer neuen Theorie des Denkens dargestellt von Salomon Maimon*. Berlin: Ernst Felisch.
Maimon, Salomon (1797): *Kritische Untersuchungen über den menschlichen Geist oder das höhere Erkenntniß – und Willensvermögen*. Leipzig: Gerhard Fischer dem Jüngern.
Maimon, Salomon (1888): *Solomon Maimon: An Autobiography* (1st ed. 1792). Murray J. Clark (Tr.). London: Alexander Gardner.
Maimon, Salomon (1966): *Giv'at Hamore* (1st ed. 1791). Samuel Hugo Bergman/Nathan Rotenstreich (Eds.). Jerusalem: The Israeli Academy of Sciences and Humanities [in Hebrew]
Maimon, Salomon (1969): "Ueber die Progressen der Philosophie veranlaßt durch die Preisfrage der königl. Akademie zu Berlin für das Jahr 1792: Was hat die Metaphysik seit Leibniz und Wolf für Progressen gemacht? (1st ed. 1793)". In: *Aetas Kantiana*. Vol. 172. Bruxelles: Culture et Civilisation.
Maimon, Salomon (1970): "Versuch einer neuen Logik oder Theorie des Denkens, Nebst angeängten Briefen des Philaletes an Aenesidemus (1st ed. 1794)". In: Valerio Verra (Ed.): *Gesammelte Werke*. Vol. 5, Hildesheim: G. Olms Verlagsbuchhandlung.
Maimon, Salomon (1971): "Maimon an Kant, 15 Mai 1790". In: Valerio Verra (Ed.): *Gesammelte Werke*. Vol. 6. Hildesheim: G. Olms Verlagsbuchhandlung, pp. 429–431.
Maimon, Salomon (1971): "Ueber den Gebrauch der Philosophie zur Erweiterung der Erkenntnis. (aus: Philosophisches Journal. 1975. Bd. II, S. 1-35)." In: Valerio Verra (Ed.): *Gesammelte Werke*. Vol. 6. Hildesheim: G. Olms Verlagsbuchhandlung, pp. 362–396.
Maimon, Salomon (1971): "Das Genie und der methodische Erfinder (aus: Berlinische Monatsschrift. 1795. BD XXVI. S. 362–384)". In: Valerio Verra (Ed.). *Gesammelte Werke*. Vol. 6. Hildesheim: G. Olms Verlagsbuchhandlung, pp. 397–420.
Maimon, Salomon (1976): "Erfindungsmethoden". In: "Ideen und Plane aus S. Ms hinterlassenen Papieren (ebenda, 1804, Heft II, S. 139–156)". In: Valerio Verra (Ed.): *Gesammelte Werke*. Vol. 7. Hildesheim: G. Olms Verlagsbuchhandlung, pp. 649–660.
Maimon, Salomon (2010): *Essay on Transcendental Philosophy* (1st ed. 1790). Midgley, Nick/Somers-Hall, Henry/Welchman, Alistair/Reglitz, Merten (Trs.). London, New York: Continuum International Publishing Group.
Maimonides (2002): *The Guide of the Perplexed*. Michael Schwarz (Tr.). Tel Aviv: Tel Aviv University Press [in Hebrew].
Manders, Kenneth (2008): "The Euclidean Diagram (1st ed. 1995)". In: Paulo Mancuso (Ed.): *The Philosophy of Mathematical Practice*. New York: Oxford University Press, pp. 80–133.
Marinos of Neapolis (1977): *The Extant Works or The Life of Proclus and the Commentary on the Dedomena of Euclid*. Alexandre N. Oikonomides (Tr.). Chicago: Ares Publishers Inc.
Martin, Gammon (1997): "Exemplary Originality: Kant on Genius and Imitation". In: *Journal of the History of Philosophy* 35. No. 4, pp. 563–592.

Melamed, Yitzhak Y. (2004): "Salomon Maimon and the Rise of Spinozism in German Idealism". In: *Journal of the History of Philosophy* 42. No. 1, pp. 67–96.
Montucla, Jean-Étienne (2007): *Histoire des mathématiques: dans laquelle on rend compte de leurs progrès depuis leur origine jusqu'à nos jours; où l'on expose le tableau et le développement des principales découvertes dans toutes les parties des mathématiques, les contestations qui se sont élevées entre les mathématiciens, et les principaux traits de la vie des plus célèbres* (1st ed. 1758). Paris: Jacques Gabay.
Morel, Thomas (2013): *Mathématiques et politiques scientifiques en Saxe (1765–1851). Institutions, acteurs et enseignements*. Doctoral dissertation. Bordeaux: Université Bordeaux 1.
Mueller, Ian (1981): *Philosophy of Mathematics and Deductive Structure in Euclid's Elements*. Cambridge, Mass.: MIT Press.
Netz, Reviel (1999): *The Shaping of Deduction in Greek Mathematics: A Study in Cognitive History. Ideas in Context* 51. Cambridge; New York: Cambridge University Press.
Newton, Isaac (1736): *The Method of Fluxions and Infinite Series: with its Application to the Geometry of Curve-lines* (1st ed. 1671). London: Henry Woodfall.
Panza, Marco (1997): "Classical Sources for the Concepts of Analysis and Synthesis", in: Michael Otte/Marco Panza (Eds.): *Analysis and Synthesis in Mathematics*. Dordrecht: Kluwer Academic Publishers, pp. 365–414.
Pappus of Alexandria (1986): *Book 7 of the 'Collection' I–II*. Alexander Jones (Ed. & Tr.). New York: Springer.
Pasini, Enrico (1997): "Arcanum Artis Inveniendi: Leibniz and Analysis". In: Michael Otte/ Marco Panza (Eds.): *Analysis and Synthesis in Mathematics*. Dordrecht: Kluwer Academic Publishers, pp. 35–46.
Philoponus (2008): *On Aristotle, Posterior Analytics 1.1–8*. Richard McKirahan (Tr.). London: Duckworth.
Popper, Karl (2005): *The Logic of Scientific Discovery* (1st ed. 1935). London; New York: Routledge.
Proclus (1789): *The Philosophical and Mathematical Commentaries of Proclus, on the First Book of Euclid's Elements, to which are added A History of the Restoration of the Platonic Theology, by the Latter Platonist. Vol. II*. Thomas Taylor (Tr.). London: Payne and Son et al.
Proclus (1992): *A Commentary on the First Book of Euclid's Elements* (1970). Glenn R. Morrow (Tr.). Princeton: Princeton University Press. (2nd ed.).
Rabouin, David (2002): *Mathesis universalis. L'idée de 'mathématique universelle' à l'âge classique*. Doctoral dissertation. Paris: Université Paris IV-Sorbonne.
Rabouin, David (2005): "La *mathématique universelle* entre mathématique et philosophie, d'Aristote à Proclus". In: *Archives de Philosophie* 68. No. 2, pp. 249–268.
Rabouin, David (2013): "*Analytica Generalissima Humanorum Cognitionum*, Some Reflections on the Relationship between Logical and Mathematical Analysis in Leibniz". In: *Studia Leibnitiana* 1. Stuttgart: Franz Steiner Verlag, pp. 109–130.
Rabouin, David (2016): "*Mathesis universalis* et algèbre générale dans les *Regulae ad directionem ingenii* de Descartes". In: *Revue d'histoire des sciences* 69. No. 2, pp. 259–309.
Reichenbach, Hans (1938): *Experience and Prediction: An Analysis of the Foundations and the Structure of Knowledge*. Chicago & London: The University of Chicago Press.

Raftopoulos, Athanassios (2003): "Cartesian Analysis and Synthesis". In: *Studies in History and Philosophy of Science* 34, pp. 265–308.
Robinson, Richard (1936): "Analysis in Greek Geometry". In: *Mind* 45. No. 180, pp. 464–473.
Rousseau, Jean-Jacques (1768): *Dictionnaire de Musique*. Paris: Veuve Duchesne.
Rubinelli, Sara (2006): "The Ancient Argumentative Game: τόποι and *loci* in Action". In: *Argumentation* 20. No. 3, pp. 253–272.
Ryle, Gilbert (1951): *The Concept of Mind* (1st ed. 1949). London: Hutchinson House.
Saito Ken/Sidoli Nathan (2010): "The Function of Diorism in Ancient Greek Analysis". In: *Historia Mathematica* 37, pp. 579–614.
Sasaki, Chikara (2003): *Descartes's Mathematical Thought*. Dordrecht; Boston: Kluwer Academic Publishers.
Schechter, Oded (2003): "The Logic of Speculative Philosophy and Skepticism in Maimon's Philosophy: Satz der Bestimmbarkeit and the Role of Synthesis". In: Gideon Freudenthal (Ed.): *Salomon Maimon: Rational Dogmatist, Empirical Skeptic: Critical Assessments*. Dordrecht; Boston: Kluwer Academic, pp. 18–53.
Schmidt, Jochen (1985): *Die Geschichte des Genie-Gedankens in der deutschen Literatur, Philosophie und Politik 1750–1945*. Vol. 1. Darmstadt: Wissenschaftliche Buchgesellschaft.
Schulz, Günter (1954): "Salomon Maimon und Goethe". In: *Goethe. Neue Folge des Jahrbuchs der Goethe-Gesellschaft* 16, pp. 272–288.
Schwab, Johann Christoph (1780): *Euklids Data, verbessert und vermehrt von Robert Simson, aus dem Englischen übersetzt und mit einer Sammlung Geometrischer, nach der Analytischen Methode der Alten aufgelößter Probleme begleitet*. Stuttgart: Christoph Freiderich Cotta.
Senderowicz, Yaron (2003): "Maimon's "Quid Facti" Argument". In: Gideon Freudenthal (Ed.): *Salomon Maimon: Rational Dogmatist, Empirical Skeptic: Critical Assessments*. Dordrecht; Boston: Kluwer Academic, pp. 176–199.
Sgarbi, Marco (2016): *Kant and Aristotle: Epistemology, Logic and Method*. Albany: State University of New York Press.
Simson, Robert (Ed. & Tr.) (1804): *The elements of Euclid, Viz. the first six books, together with the eleventh and twelfth. The errors, by which Theon, or others, have long ago vitiated these books, are corrected, and some of Euclid's demonstrations are restored. Also, the book of Euclid's Data, in like manner corrected* (1st ed. 1756). London: Wingrave.
Simson, Robert (Ed. & Tr.) (1811): *The elements of Euclid, Viz. the first six books, together with the eleventh and twelfth. The errors, by which Theon, or others, have long ago vitiated these books, are corrected, and some of Euclid's demonstrations are restored. Also, the book of Euclid's Data, in like manner corrected* (1st ed. 1756). Philadelphia: Johnson and Warner.
Sinaceur, Horuria Benis (1989): "Ars inveniendi, aujourd'hui". In: *Les Études philosophiques* 2, pp. 201–214.
Socher, Abraham, P. (2006): *The Radical Enlightenment of Solomon Maimon: Judaism, Heresy, and Philosophy*. Stanford, California: Stanford University Press.
Spranzi, Marta (2011): *The Art of Dialectic between Dialogue and Rhetoric: The Aristotelian Tradition*. Amsterdam/Philadelphia: John Benjamins Publishing Company.

Sylvain, Zac (1986): "Salomon Maïmon et les malentendus du langage". In: *Revue de Métaphysique et de Morale* 91. No. 2, pp. 181–202.
Taisbak, Christian Marinus (1991): "Elements of Euclid's *Data*". In: *Apeiron* 24. No. 4, pp. 135–171.
Taisbak, Christian Marinus, (Ed. & Tr.) (2003): *ΔΕΔΟΜΕΝΑ: Euclid's Data or the Importance of being Given*. University of Copenhagen: Museum Tusculanum Press.
Thielke, Peter (2015): "The Spinozistic Path to Skepticism: Maimon, Novalis, and the Demands of Reason". In: *Idealistic Studies* 45. No. 1, pp, 1–19.
Timmermans, Benoît (1993): "The Originality of Descartes' Conception of Analysis as Discovery". In: *Journal of the History of Ideas* 60. No. 3, pp. 433–447.
Tonelli, Giorgio (1966): "Kant's Early Theory of Genius (1770–1779): Part I". In: *Journal of the History of Philosophy* 4. No. 2, pp. 109–132.
Ungar, Abraham A. (1999): "The Hyperbolic Pythagorean Theorem in the Poincare Disc Model of Hyperbolic Geometry". In: *The American Mathematical Monthly* 106, No. 8, pp. 759–763.
Van Peursen, Cornelis Anthonie (1986): "Ars Inveniendi bei Leibniz". In: *Studia Leibnitiana* 18. No. 2, pp. 183–194.
Walker, Alan (2001): "Indexing Commonplace Books: John Locke's Method". In: *The Indexer* 22. No. 3, pp. 114–118.
Wallace, Karl R. (1973): "Francis Bacon and Method: Theory and Practice". In: *Speech Monographs* 40. No. 4, pp. 243–272.
Wolff, Christian von (1734): *Vollständiges mathematisches Lexicon*. Leipzig: Friedrich Gleditschens.
Yakira, Elhanan (2003): "From Kant to Leibniz? Salomon Maimon and the Question of Predication". In: Gideon Freudenthal (Ed.): *Salomon Maimon: Rational Dogmatist, Empirical Skeptic: Critical Assessments*. Dordrecht; Boston: Kluwer Academic, pp. 54–79.
Zahar, Elie (1983): "Logic of Discovery or Psychology of Invention?". In: *The British Journal for the Philosophy of Science* 34. No. 3, pp. 243–261.
Zheng, Fanglei (2012): *Des Data d'Euclide au De numeris datis de Jordanus de Nemore – Les Données, l'Analyse et les Problèmes*. Doctoral dissertation. Paris: Université Paris 7 – Denis Diderot.
Zompetti, Joseph P. (2006): "The Value of Topoi". In: *Argumentation* 20. No. 1, pp. 15–28.

Index of Terms

a posteriori 25, 31, 62, 73, 151
a priori 4, 25, 28 f., 44, 51 f., 62, 64, 71–76, 78, 149–152
abduction 60
abstraction 101, 140
actual 26, 34 f., 58, 81 f., 115, 121, 144, 147 f.
actuality 4, 37–39, 44 f., 69, 71, 78, 111, 114 f., 148
aesthetic 6, 16, 18
algebra 24, 28, 30, 34–37, 39, 41, 50, 115, 144
algebraic geometry 35
ampliative analysis 4, 47, 60, 73, 76 f., 150
analogy 29
analysis in the broader sense 4, 47, 49, 59, 150–152
analysis in the narrow sense 4, 73, 150
analysis of conditions 57 f., 84 f., 93 f., 97, 100
analysis of the concept 59, 65, 101
analysis of the object 4, 30, 39, 47, 57, 59–68, 71 f., 76–78, 81, 107, 109–112, 128, 146, 149, 151
analysis situs 34, 39 f.
apodictic 74–76, 152
arithmetic 9, 28, 34, 38, 40, 74, 97
ars characteristica 33 f., 148
ars combinatoria 33
ars demonstrandi 36
ars inveniendi 1, 4, 27, 30 f., 33, 35–37, 47, 83, 148
art of argumentation 32
art of discovery 34, 47, 50
art of finding arguments 27, 31 f., 34, 39, 145, 148
art of thinking 34 f.
assertoric 74 f., 152
attribute 1 f., 12, 17, 28, 52, 66, 76, 78 f., 114, 132, 137 f., 144, 147, 150, 152 f.
axiom 52, 73, 98 f., 124, 135 f., 142

beautiful 16 f., 26
beauty 16, 20

calculation 24, 141
calculus 14, 22, 24, 37, 39 f., 148
categorical form 126
certainty 25, 35
certitude 27, 36
chance 7, 16, 23, 25, 27 f., 134 f.
cognition 8, 10, 12, 19, 23, 29 f., 35 f., 38 f., 41, 43–45, 50, 52, 66, 68 f., 71–73, 78, 129, 147–151
concept 1, 4, 6 f., 12–17, 19, 26, 29 f., 36, 38–40, 43 f., 46, , 51 f., 55–65, 69–72, 74–76, 78 f., 101 f., 120 f., 125 f., 130, 136, 147, 149–153
conditio sine qua non 56, 125 f.
condition 27, 32, 44, 46, 49, 54, 57–59, 66, 81, 84–97, 99–102, 104, 106 f., 112–115, 119, 125, 128, 132, 138 f., 150
consequence 48, 55, 57, 63, 66, 69–71, 114, 143, 145, 151
construction 9, 20, 22, 29 f., 33, 38, 40, 48 f., 51–53, 58 f., 61, 63, 66 f., 71, 78, 87, 89, 97–99, 102, 107, 109–115, 117, 121–123, 128, 131, 146, 149, 151
context of justification 9
continuity 39
converse 132–139, 143
conversion 1, 68, 132–137, 140 f., 148
corollary 134 f., 140
cosine 38
culture 10, 35

deception 85
demonstration 22, 39, 44, 48, 50, 52, 56, 61, 65, 67 f., 81, 109, 112 f., 125, 131, 135, 142–144, 148 f., 151
determinable 39, 55, 63, 65 f., 69–71, 73–75, 78 f., 129–131, 144, 151–153
determinant 52, 55, 62–65, 69, 71, 73–75, 78 f., 129 f., 152 f.
determination 40, 46, 64–66, 69–71, 76, 78 f., 85, 109, 116–118, 130, 135, 151–153
differential 14, 37, 39 f.

diorism 57 f., 83, 150
discovery 3 f., 7, 9, 20, 26–29, 31, 33 f., 36, 47, 50, 57, 76 f., 79, 136, 150, 152

Enlightenment 15, 36
ens imaginarium 52
ens rationis 51
equation 28, 36, 115, 144
experience 1, 62, 73, 78, 147, 150

faculty 1 f., 5, 7 f., 15, 17, 20, 25, 32, 38, 42–45, 64, 67–69, 78, 111, 120, 129, 137, 148, 151
fictions 37–39, 44, 68
fine art 16–19
fluent 14
fluxion 14

generalization 68, 102, 132, 137–141
Geniezeit 3, 6, 11 f., 26
genius 2 f., 6–21, 25 f., 28, 30, 33, 47, 112, 120, 147
– scientific genius 2, 17 f., 20, 147
geometry 4, 8 f., 18, 24, 28, 33 f., 39 f., 50, 56, 97, 115, 119, 148
– analytical geometry 37, 41, 144
– geometrical analysis 48
– Euclidean geometry 2, 4., 32, 34–36, 39 f., 84, 107, 118, 123 f., 134, 143–145, 148 f., 153
– Greek geometry 32 f., 36 f., 113
– hyperbolic geometry 123–124
– pure geometry 40
given as a proposition 42 f., 149
given in actuality 37–39, 45, 69, 78, 111, 148
given in form 77, 86, 97–100
given in intuition 29, 37 f., 41–46, 69 f., 73, 111, 121, 128, 148–151
given in magnitude 45 f., 77, 86, 97–101, 110 f., 113, 121, 126–128, 131, 150
given in position 45, 86, 98–101, 110 f., 116 f., 121–123
given in the broader sense 41–43, 150
given in the narrow sense 41–43, 149
givenness 45, 85 f., 90, 94, 123, 128
God 1, 8, 23, 42, 78, 153

heuresis 31
hypothetical form 54 f., 126, 131, 134

identity 39, 46, 50 f., 60, 63, 73, 76, 128, 143–146, 148 f.
imagination 15–17, 34, 38, 42–44, 63 f., 68, 71, 111, 120, 147 f., 151
imitation 11–13, 17
imitator 3, 11, 19 f., 22 f.
indivisible 14, 28, 37
inference 52, 60, 62, 66, 68, 103, 111, 121, 124, 128 f., 131, 134, 137, 151
infinite 14, 38, 69, 86–89, 91, 98–101, 123
infinitely small 14, 38 f.
infinitesimal 14, 41
inspiration 8, 16
instrument 49, 83
intuition 3 f., 8, 16, 37–40, 42–45, 51 f., 59, 62, 66, 69–71, 73, 75, 78, 86, 111, 114, 147–151, 153
inventive power 20

judgment 4, 9 f., 13, 15 f., 39 f., 43, 50, 52–55, 58, 60, 62–68, 71–77, 79, 85, 120, 125 f., 137, 144–146, 148–152

Kabbalah 35
knowledge 1–5, 7 f., 10, 12, 17–19, 22–24, 26–28, 30, 32, 34–36, 38–44, 49–51, 55 f., 60, 62–66, 71 f., 76, 78, 80–82, 86, 96, 118–120, 129, 147, 150, 152 f.
– analytic knowledge 60, 65, 73, 141
– certain knowledge 30
– synthetic knowledge 40, 65, 68, 73, 78, 102, 141

language 13, 23, 36, 38, 78
law 7 f., 11, 15, 28 f., 32, 41 f., 44, 85, 120, 124
locus, loci 31, 34
logic 3 f., 15, 23, 30, 34, 39, 47, 49–51, 54, 58–63, 67, 71, 73, 75, 77, 83, 111, 120 f., 125, 128–130, 146–150
logical analysis 3 f., 15, 30, 41, 47, 49–53, 55 f., 84, 111, 124–126, 128, 132, 143, 146, 148–151

Index of Terms — 165

machine   23, 28 f.
magnitude   14, 30, 33, 39 f., 45 – 46, 59, 84, 86, 98 – 101, 111, 116 f., 122, 124 – 128, 130 f., 149
manifold   40, 42 f., 66, 72, 150 f.
mathematical analysis   5, 50 – 52, 120 f., 143, 150
mathesis universalis   19, 30
memory   34, 42, 52
metaphysics   35, 37, 40 f.
methodical inventor   2 f., 6 – 11, 17 – 21, 23, 25 f., 30, 47, 80, 97, 104, 112, 120, 133, 147
middle term   8, 29, 58 f., 129 – 131
modality   54, 74

natural   13, 15, 18, 23, 25 f., 28 f., 48, 85, 121, 124
nature   10, 12 f., 15 – 19, 26, 44, 58
necessity   29, 32 f., 50, 62 – 65, 72 f., 75, 78, 148, 151
nihil negativum   51
nihil privativum   51
nominal definition   65
notation   40, 129

objective   7, 10, 19, 25 f., 51, 62 – 64, 75, 77 f., 144, 151
observation   21, 24, 121, 143
originality   3, 11 – 15, 17

perception   21, 143
phenomenon, phenomena   1, 23, 28 f.
plurality   2 f., 44
poetry   8 f., 18 f.
porism   32 f.
possibility   4, 29., 33, 35, 37, 41, 51, 72 – 75, 86, 89 f., 96, 107, 114 f., 119, 138, 142, 152
possible   1, 3, 5, 8, 10, 16 f., 26, 28, 33, 37, 39 – 42, 45, 49 – 51, 55, 57 f., 63 – 65, 68 f., 71 f., 74 f., 78 f., 81, 84 – 86, 90, 92 f., 98 f., 106 f., 110, 112 – 115, 118 – 121, 123 f., 141 – 143, 148 f., 152 f.
predicate   4, 33, 38, 40, 43, 51 – 53, 55, 58, 60 – 67, 69 – 72, 74 – 76, 129 f., 133 f., 136 f., 141, 148, 150 – 152

premise   3, 8, 27, 29 – 32, 37, 41 f., 44, 47, 50 f., 56 f., 62, 83, 103, 107, 118, 126 – 129, 131, 137, 142 f., 145, 147 – 149
pretense   85
principle   4 f., 18, 23 f., 41, 45, 48 – 50, 67 – 69, 73, 110, 128, 131, 144, 151
– logical principle   50, 53, 121, 153
– principle of contradiction   3 f., 50 – 51, 53, 55, 59, 62 f., 66, 72 f., 75, 101, 120, 125, 128, 132, 144, 149 – 151
– principle of determinability   4, 40 f., 55, 63, 65 f., 69, 129 f., 151
property   12, 15, 33, 61 f., 64 f., 67, 70, 76, 87, 114, 152
prophecy   15
proportion   22, 30, 33, 55, 59, 77, 135
pure   9, 17, 24, 32, 34, 38, 40, 72 f., 78, 80, 111, 121, 150, 152
– pure time   40

quality   30, 40, 54, 83
quantity, *quantum*   30, 40, 54, 127, 136

ratio   39, 46, 58, 78, 86, 124 – 128, 130, 135
rationality   1
real   1, 12, 14, 18, 26, 29, 37 f., 51, 65 f., 69, 71, 73, 77 – 79, 115, 120 f., 143, 148, 152 f.
– real definition   51
– real synthesis   66, 69 – 71, 74, 76
– real thought   69, 71, 151
reality   19, 25 f., 51, 69
reductio ad absurdum   70, 121 f., 133, 142, 145
regressive analysis   125 f., 128, 132, 149
relation   3, 5, 7, 15, 30, 33 – 40, 43 – 47, 51, 54 f., 58 – 60, 63, 65 f., 71 – 76, 78, 96, 101, 106 f., 124, 127, 131, 142, 149, 151 f.
– relation of determinability   4, 41, 55, 66, 70 f., 74, 129, 131, 151
representation   37, 40, 42 f., 73, 85 f.
rhetoric   31, 34 – 36, 41, 82, 84
rigor   22, 25

schema   79
science   3, 5, 9, 11, 16–20, 22–26, 28–30, 34–37, 39–40, 50, 52, 80, 82, 95. 120f., 128, 148
– general science   34–36
– science of figures   39f., 148
– science of space   39f.
scientist   18, 27
sensibility   42, 44, 73, 86, 119, 147, 149
similitude   30
space   37, 39f., 42–44, 61, 63, 65, 70f., 73, 79, 126–128, 131, 148, 153
syllogism   3, 27, 29–32, 35, 37, 41, 44f., 47, 49–51, 56, 68, 121, 125–132, 137, 145, 147–150
synthesis   1, 3f., 33, 41, 44, 47–49, 51, 61f., 65f., 68–72, 75–79, 83, 129f., 140–147, 149–152
system   2, 21, 24, 27, 33, 36, 39, 52, 55f., 66, 82, 121, 123f., 153
– coalition-system   24

talent   10f., 15, 18f.
taste   6, 16f.
technique   6, 31, 47, 83

theorem   8f., 27, 32f., 92, 106, 110, 112, 114–118, 124, 131, 135, 142, 145
– Pythagorean Theorem   54, 61., 65–67, 118, 123f., 134, 139f., 151
thinking   24, 43, 45, 55, 60f., 70, 78f., 130, 144, 153
thought   1, 9, 16, 19–21, 38, 40, 42–45, 50, 55f., 58, 60f., 63–66, 69–74, 76, 78f., 119, 130, 144, 152f.
– analytic thought   61f., 66, 76, 78
– arbitrary thought   69
– synthetic thought   61, 73, 76, 78
topics, *topoi*   31f., 34f., 82
truth   1f., 5, 8, 11f., 20–25, 27–29, 34f., 38, 41, 47f., 50, 52, 54, 57, 60, 62, 77, 80, 82, 103, 118–121, 123f., 131f., 142–144, 146–149, 153

understanding   5, 15f., 24, 39, 43f., 49, 61, 63, 67f., 71, 73, 78f., 83, 103, 117, 119f., 137, 147–149, 151f.
– finite understanding   1, 75f., 78f., 152f.
– infinite understanding   1, 76, 78f., 153
unit   33, 38, 68, 150
unity   40, 43f., 66, 72, 78, 150f.
universality   113, 137–139
universalization   102, 140f.

Index of Person

Alexander of Aphrodisias 31f., 131, 133, 145
Apollonius 32, 58, 91, 107
Archimedes 23
Aristotle 31–33, 62, 66, 82, 131, 133, 136, 138, 145, 148
Arnauld, Antoine 34

Bacon, Francis 20f., 23, 31, 34, 82f., 147
Batteux, Charles 13, 17, 147
Ben Makhir (ibn Tibbon), Jacob 45
Buhle, Theophilus 32

Cardano, Girolamo 28
Cavalieri, Bonaventura 14
Cicero 31, 34, 36
Clavius, Christopher 81f., 119, 137-139

D'Alembert, Jean le Rond 12, 27f., 132, 137
De Saint-Lambert Jean François 12
Del Ferro, Scipione 28
Descartes, René 2, 19, 22, 28, 30f., 33, 47, 81-83, 85, 93, 115, 118, 147f.
Diderot, Denis 12, 27

Euclid 2, 9, 23, 32-34, 45, 48, 56, 70, 77, 81-83, 86f., 89-91, 94f., 99, 102-105, 107f., 112, 115-119, 121, 123f., 126f., 130f., 133, 142, 145, 150
Euler, Leonhard 36

Fichte, Johann Golttlieb 25f.
Flögels, Carl Friedrich 35

Garve, Christian 17
Gerard, Alexander 17, 20, 147
Goethe, Johann Wolfgang 2, 35
Grothendieck, Alexander 35
Guericke, Otto von 29, 147

Heron of Alexandria 103f., 119
Hilbert, David 135
Homer 19

Hume, David 24
Huygens, Christian 29

Ibn Ishaq, Hunayn 45

Kant, Immanuel 4, 10, 12f., 15-21, 23f., 34, 44, 49-51, 54f., 60, 65f., 71-76, 78, 86, 120, 147, 152
Kästner, Abraham Gotthelf 24, 97
Kepler, Johannes 29
Klopstock, Friedrich Gottlieb 9, 19

Lambert, Johann Heinrich 12, 29, 31, 50
Leibniz, Gottfried Wilhelm 2, 14, 21–24, 27, 30f., 33–37, 39f., 47f., 50, 52–54, 57, 66, 102, 140-141, 143, 147f.
Lessing, Gotthold Ephraim 17
Locke, John 36, 101
Lorenz, Johann Friedrich 81, 86, 97, 110
Lucretius 9

Mach, Ernst 10
Maimonides 78, 119, 133
Marinos of Neapolis 30, 83
Montucla, Jean-Étienne 21, 142

Napier, John 28
Newton, Isaac 8–10, 14, 18, 20, 28
Nicole, Pierre 34
Nieuwentijt, Bernard 14
Novalis 19

Pappus of Alexandria 32f., 35, 45, 48, 58, 83, 91, 134, 139
Peletier, Jacques 35, 135
Philoponus 138
Plato 48
Porphyry 48
Proclus 4, 33, 48f., 57, 82, 88, 90-92, 96, 104-106, 109f., 113f., 119, 133, 135, 138, 142, 148
Pyrrho 21

Reichenbach, Hans 9
Rousseau, Jean-Jacques 14f.
Russell, Bertrand 135
Ryle, Gilbert 6

Schlegel, Johann Adolf 13
Schwab, Johann Christoph 2, 45, 56–59, 64, 81, 125-128, 142
Segner, Jan Andrej 36
Simson, Robert 46, 81, 85f., 89f., 97, 106, 110, 119, 122f., 125-127, 130f., 145
Spinoza, Baruch 24

Thales 21
Torricelli, Evangelista 14

Von Tschirnhaus, Ehrenfried Walther 28

Wallis, John 29
Wieland, Christoph Martin 19
Wolff, Christian von 30, 33, 35, 37, 62, 131
Wren, Christopher 29

Xenophon 21

Young, Eduard 13

Zeno 96

www.ingramcontent.com/pod-product-compliance
Lightning Source LLC
Chambersburg PA
CBHW020332170426
43200CB00006B/357